About Professional Baking
Student Workbook

About
Professional Baking
Student Workbook

THOMSON

DELMAR LEARNING

Australia Canada Mexico Singapore Spain United Kingdom United States

© 2006 by Thomson Delmar Learning, a part of the Thomson Corporation. Thomson, the Star logo, and Delmar Learning are trademarks used herein under license.

Printed in the United States of America
1 2 3 4 5 XXX 10 09 08 07 06

For more information contact Thomson Delmar Learning,
Executive Woods, 5 Maxwell Drive, Clifton Park, NY 12065-2919

Or find us on the World Wide Web at http://www.delmarlearning.com or www.culinary.delmar.com

ISBN: 1-4180-1972-0

CONTENTS

INTRODUCTION

The material in this workbook is designed to reinforce the information in *About Professional Baking,* by Gail Sokol. Each chapter in this workbook, as in the book, begins with objectives. After reading the chapter and completing the material in this workbook, you should reach the goals that are outlined. If not, you should reread the material and speak with your instructor about any problem areas.

This workbook also includes a brief outline of the material in each chapter. The outlines are presented as refreshers and can help put the important features of the material into perspective.

The key terms that you see in both the book and the workbook are important to master. Some terms that are especially important are repeated. Seeing the same word, such as gluten, several times should indicate to you that it is essential to the baking process.

The workbook, as does the book, uses a building block approach so that information in one chapter often builds on what was presented in the one before. For this reason it is important to fully understand the terms and concepts in one chapter before moving on to the next.

The questions are in common and easily understood formats—fill in the blanks, true or false, multiple choice, and matching. Keep in mind that even though answering the questions correctly is a goal, it is also important to know why the answers are correct.

This workbook is, above all, meant to be an aid in your study and application of the principles covered in *About Professional Baking*.

To view the Online Companion,™ please visit www.culinary.delmar.com.

CHAPTER 1

Mise en Place

After reading this chapter and successfully answering the following questions, you should be able to:

■ Demonstrate organizational skills necessary to bake successfully.

■ Demonstrate proper measurement of solids and liquids by volume.

■ Demonstrate how to weigh ingredients on a baker's scale and a digital scale.

A Brief Outline of the Chapter

I. Organizational skills needed for baking

   A. Reading the recipe carefully

   B. Identifying tools and/or equipment needed

   C. Gathering all necessary ingredients, tools, and equipment close to the work area

   D. Preparing the pans

   E. Measuring properly

   F. Practicing good sanitation

      1. Temperature danger zone

      2. Safe handling of eggs

      3. Safe handling of dairy products

II. How to measure the ingredients in this book

   A. By weight

   B. By volume

   C. By length

   D. By temperature

III. The correct tools of measurement

    A. Properly measuring a liquid by volume

    B. Properly measuring of a solid by volume

IV. How to use a baker's scale

V. How to use a digital scale

VI. Baker's percentages

VII. Baking tools and equipment defined

VIII. Procedures for preparing pans for baking

    A. Procedure to make parchment cake pan liners for round cake pans

    B. How to line sheet pans with parchment paper

    C. Proper preparation of cake pans before baking

    D. Preparing muffin tins

IX. Forming a parchment cone

X. Sifting dry ingredients

XI. Storing baked goods

XII. Properly wrapping baked goods to be frozen

Key Terms

Explain these key terms.

1. baker's percentages

2. calibrated

3. formulas

4. meniscus

5. mise en place

6. scaling

7. temperature danger zone

Fill in the Blanks

Use the most accurate word or phrase to complete each sentence.

8. The six main organizational skills needed for baking include reading the recipe carefully; identifying tools and equipment that will be needed; gathering all the ingredients, tools, and equipment close to the work area; preparing pans; measuring ingredients; and _____ _____.

9. The French term _____ is used by chefs for the act of getting all the ingredients, equipment, and tools in place before baking.

10. Professional bakers refer to recipes as _____.

11. Reading the recipe carefully assures that the proper ingredients, equipment, and tools will be available; that the baker can envision exactly what each stage of the process will look like; and that the baker can identify the appropriate _____.

12. The term _____ is used when a baker divides a piece of dough into specific amounts.

13. Microorganisms that can multiply in food and ultimately cause sickness in humans are called _____.

14. Microorganisms that can be carried in dairy products include Campylobacter jejuni, _____, and Listeria monocytogenes.

15. The system for measuring used in Europe and other parts of the world is known as the _____ system.

16. A baker measures ingredients by weight using either a digital scale or a _____.

17. The two ways to measure ingredients by volume is to use _____ for liquids and _____ to measure solids.

18. It is best to zero a baker's scale or _____ it before weighing anything to make sure that it is balanced properly.

19. In a baker's formula for white bread, the _____ is usually equated to 100 percent.

20. _____ is a French term for a warm water bath.

True or False

Identify each statement as true (T) or false (F).

_____ 21. Mise en place is done only by bread bakers.

_____ 22. Fluid ounces and dry ounces will always be the same.

_____ 23. Professional bakers tend to measure ingredients by weight.

_____ 24. In the U.S. system, weight is measured in pounds and ounces.

_____ 25. In the metric system, weight is measured in liters and milliliters.

_____ 26. Eggs may be measured by weight or volume.

_____ 27. One disadvantage in using a baker's scale is that a breeze could throw off the balance.

_____ 28. Parchment paper will burn in a hot oven and must be used at low to moderate temperatures only.

_____ 29. A bain marie will help to rapidly chill an ice cream.

_____ 30. When preparing the pans for a multilayer cake, each pan should be lined with parchment circles.

Multiple Choice

Identify the choice (a, b, or c) that most accurately answers the question or completes the statement.

31. Which of the following microorganisms can be found in dairy products?
    a. *Salmonella*
    b. hepatitis B
    c. trichinosis

32. Milk and dairy products should be stored at
    a. 41°F or higher
    b. 45°F to 50°F
    c. 41°F or lower

33. The danger zone for hot foods is
    a. 41°F or lower
    b. 41°F to 135°F
    c. 135°F or higher

34. The following are good tips for the safe handling of eggs except
    a. always keep eggs in their original container
    b. wash your hands before and after handling eggs
    c. serve raw eggs when called for in a recipe

35. Measuring liquid by volume should be done
    a. by placing the measuring cup on a level surface
    b. by carefully holding the cup up to your eye
    c. by always filling it to the rim

36. The first step in using a baker's scale is to
    a. add ingredients to the scoop or container
    b. tare or zero the scale
    c. set the desired weight

37. When a baker scales the ingredients, it means
    a. the baker is weighing the ingredients
    b. the baker is kneading the dough
    c. the baker is determining doneness

38. The following are examples of safe sanitation practices except
    a. washing your hands
    b. keeping the dairy ingredients at a temperature of 51°F
    c. keeping eggs in their original container

39. Cross contamination occurs
    a. when bacteria are spread from one surface to another
    b. when the temperature falls below the danger zone
    c. when raw eggs are consumed

40. In the United States, temperature is measured in Fahrenheit, whereas in Europe and other parts of the world, it is measured in
    a. meters
    b. liters
    c. Celsius

Matching

Match each term in column 1 with its definition from column 2.

_____ 41. baguette pan
_____ 42. bain marie
_____ 43. bench scraper
_____ 44. double boiler
_____ 45. false bottom tart pan
_____ 46. parchment paper
_____ 47. ramekin
_____ 48. silicone baking mat
_____ 49. spring form pan
_____ 50. coupler

a. pan used for making fluted pastry crust

b. used to line cake and sheet pans

c. small baking dish usually made of ceramic

d. mat able to withstand high temperatures

e. round pan with removable bottom

f. double pot with water on bottom

g. French term for warm water bath

h. a long metal pan formed in to a loaf cylinder

i. small rectangle blade to cut and scale

j. allows tips to fit into pastry bag

CHAPTER 2

Ingredients

After reading this chapter and successfully answering the following questions, you should be able to:

- Define the function of each of the major ingredients used in the bakeshop.

- Describe how ingredients work together within recipes to produce a specific finished product.

A Brief Outline of the Chapter

B. Coconut

C. Toasting coconut and nuts

XII. Fruits

Key Terms

Explain these key terms.

1. dry ingredients

2. homogenization

3. pasteurization

4. plasticity

5. solid ingredients

6. ultrapasteurization

7. wet ingredients

Fill in the Blanks

Use the most accurate word or phrase to complete each sentence.

8. There are basically two types of wheat: _____ and
    _____ .

9. The energy nutrient group to which sugar belongs is _____ .

10. The group of ingredients that are usually used to thicken things, such as
    pie fillings or sauces, is _____ .

11. _____ is the category of ingredients used to stabilize desserts.

12. The group of ingredients also known as lipids is _____ and are
    not soluble in water.

13. The plasticity of a fat is determined by the _____
    and the surrounding temperature.

14. _____ oil is high in mono-unsaturated fats and is made from
    rapeseed.

15. Homogenization is a process to distribute the _____ through-
    out the milk.

16. Low-fat milk can contain from _____ to _____ percent
    milk fat.

17. The deep flavor imparted by spices is derived from the _____
    that they contain.

18. Extracts and emulsions, whether natural or artificial, are examples of
    _____ .

19. The two common types of salt used in the text are table salt, which is a
    fine grain crystal, and _____ salt, which has a coarser grain.

20. Nut flours often have _____ added to absorb the natural oils
    from the nuts.

True or False

Identify each statement as true (T) or false (F).

_____ 21. Unless the product is crystal in form and white, it cannot be classified as a sugar.

_____ 22. Fresh whole milk has approximately 3½ percent milkfat.

_____ 23. Heavy cream contains at least 50 percent milkfat.

_____ 24. Nut butters usually have a small amount of oil added to improve the smooth texture.

_____ 25. Marzipan is made from any nut butter.

_____ 26. Cream of coconut usually has sugar added in the processing.

_____ 27. Vegetable shortenings began as oils and then were hydrogenated to turn the oils into a solid form.

_____ 28. Gelatins may be either animal or plant proteins.

_____ 29. Hydrogenating oils increases their shelf life.

_____ 30. Praline can be ground and made into a paste form.

Multiple Choice

Identify the choice (a, b, or c) that most accurately answers the question or completes the statement.

31. Brown sugar is granulated sugar with
    a. honey added
    b. molasses added
    c. confectioner's sugar added

32. Sugar mixed with water will prevent the water from
    a. freezing solid
    b. becoming an acid
    c. becoming an alkali

33. Fat performs the following functions in baked goods, except
    a. adds flavor
    b. acts as a tenderizer
    c. decreases the shelf life

34. Because of their high degree of plasticity, high-ratio or emulsified short-enings should not be used when
    a. baking cakes
    b. a thin batter is needed
    c. the creaming method of mixing is called for

35. Skim milk contains
    a. no milk fat
    b. between 2 percent and 3 percent milkfat
    c. up to ½ percent milkfat

36. Vanilla beans are the dried fruit of a
    a. tropical orchid plant
    b. tropical palm tree
    c. tropical lily plant

37. When ground, it is mace; but when in seed form, it is
    a. nutmeg
    b. cloves
    c. ginger

38. The flavor imparted by the litchi, or lychee, would best be described as
    a. tart
    b. lightly sweet
    c. pungent

39. Of the following nuts, the one that must be cooked before eating is the
    a. chestnut
    b. walnut
    c. almond

40. Peanuts are not true nuts but rather
    a. legumes
    b. fruits
    c. vegetables

Matching

Match each term in column 1 with its definition from column 2.

_____ 41. allspice
_____ 42. cinnamon
_____ 43. clove
_____ 44. ginger
_____ 45. nutmeg
_____ 46. saffron
_____ 47. vanilla
_____ 48. marzipan
_____ 49. gianduja
_____ 50. nut flour

a. inner bark of an evergreen laurel tree

b. dried, opened flower buds of the tropical evergreen tree

c. tropical plant native to China

d. hard seed from the tropical evergreen tree

e. inner threads (stigma) inside the purple crocus plant

f. from plant with white flowers and reddish brown berries

g. dried fruit of a tropical orchid

h. combination of hazelnuts and chocolate

i. finely ground nuts

j. almond paste and sugar with dough-like consistency

CHAPTER 3

Wheat Flour— The Essential Grain (and Other Flours)

After reading this chapter and successfully answering the following questions, you should be able to:

- List the five stages of the milling process.

- Explain the importance of wheat flour to the baker.

- Define how proteins in wheat flour relate to gluten.

- Understand the difference between hard wheats and soft wheats.

- Understand when to use a low-protein flour.

- Understand when to use a high-protein flour.

- Explain how flour is treated after milling.

- Understand the importance of gluten in making yeast breads.

- Control gluten in baked goods without yeast.

A Brief Outline of the Chapter

Key Terms

Explain these key terms.

1. all-purpose flour

2. bleaching

3. bread flour

4. breaking

5. cake flour

6. classifying

7. clear flour

8. gliadin

9. gluten

10. glutenin

11. pastry flour

12. patent flour

13. purifying

14. reducing

15. sifting

16. streams

Fill in the Blanks

Use the most accurate word or phrase to complete each sentence.

17.   The outer shell of a wheat kernel is called _____.

18.   The germ is the _____ part of the wheat kernel.

19.   The gluten-forming proteins are found in the _____ of the wheat.

20.   The eleven varieties of wheat flour include whole wheat, patent, bread, hard patent, clear, straight, high-gluten, patent durum, all-purpose, _____, and _____.

21.   The protein content of high gluten flour is _____, whereas cake flour is _____.

22.   The term for the whole wheat kernel before processing is a _____.

23.   _____ flour contains the least bran and germ particles.

24.   Whole wheat flour is made from the entire wheat kernel, the endosperm, bran, and _____.

25.   The hardest wheat grown is called _____.

True or False

Identify each statement as true (T) or false (F).

_____ 26. Rye flour is good for baking breads because it is very good in forming gluten.

_____ 27. Buckwheat is a wild form of wheat.

_____ 28. Gluten is a combination of gliadin and glutenin.

_____ 29. Starch is a protein that makes up three quarters of the wheat grain.

_____ 30. A cake flour would have a higher amount of protein than an all-purpose flour.

_____ 31. Pumpernickel flour is another name for whole rye flour.

_____ 32. A medium rye is higher in protein than a light rye.

_____ 33. Bleaching reduces the quality of gluten formation in a flour.

_____ 34. Bromated flour refers to the addition of potassium bromate.

_____ 35. Starch makes up less than one half of the wheat grain.

Multiple Choice

Identify the choice (a, b, or c) that most accurately answers the question or completes the statement.

36. The following are often added to enriched flour except
    a. iron
    b. vitamin B
    c. vitamin E

37. The ash content of white flour versus whole wheat would be
    a. higher
    b. about the same
    c. lower

38. Harder wheat flours versus softer wheat flours
    a. form more gluten
    b. form less gluten
    c. form no gluten

39. One way to control or lessen gluten formation in a recipe is to
    a. increase the fat
    b. decrease the fat
    c. increase the mixing time

40. The exposure of milled flour to chlorine gas or benzoyl peroxide is a process called
    a. bromating
    b. bleaching
    c. enriching

Matching

Match each term in column 1 with its definition from column 2.

_____ 41. high-gluten flour

_____ 42. durum flour

_____ 43. clear flour

_____ 44. soft wheat

_____ 45. all-purpose flour

_____ 46. cake flour

_____ 47. gluten

_____ 48. wheat berries

_____ 49. bran

_____ 50. germ

a. also called semolina flour

b. used in bread dough for a chewier texture

c. combination of hard and soft wheat

d. smallest part of the wheat kernel

e. hull of the wheat kernel

f. unprocessed whole wheat kernel

g. comprised of gliadin and glutenin

h. very low protein flour

i. remains after patent flour is removed

j. red and white varieties

CHAPTER 4

The Science of Mixing

After reading this chapter and successfully answering the following questions, you should be able to:

- Understand the four mixing factors that affect baked goods.

- Understand the three main reasons for mixing baked goods.

- Explain what gluten is.

- Explain how mixing affects gluten development.

- Explain how fats play a role in mixing.

- List the correct tools for mixing.

- Properly fold in ingredients.

A Brief Outline of the Chapter

V.   Five of the most common mixing terms

VI.   Properly folding in ingredients

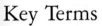

Key Terms

Explain these key terms.

1.   dough hook

2.   fats

3.   folding

4.   food processor

5.   gluten

6.   homogenous

7.   paddle

8. pastry blender

9. rubber spatula

10. whip

11. whisk

_____

Fill in the Blanks

Use the most accurate word or phrase to complete each sentence.

12. Mixing helps dry ingredients break down so that more _____ can be absorbed.

13. Incorporating molecules of air through mixing adds _____ to the texture of the finished product.

14. Overmixing can develop too much gluten and thus cause _____ in baked goods.

15. When folding meringue into a batter, you should use a _____.

16. The _____ of mixing helps dry ingredients break down into finer particles.

17. In a yeast bread, you would want to develop _____ gluten, but in a flaky pie crust, you would want to develop _____ gluten.

18. Fats aid the _____ process with the incorporation of air through mixing.

19. During the process of kneading, _____ gas is trapped, which helps to leaven the bread.

20. The method usually used to add beaten egg whites into a batter is _____.

True or False

Identify each statement as true (T) or false (F).

_____ 21. One unwanted outcome of undermixing is less leavening power.

_____ 22. The water is always added with the flour in making a flaky pie crust dough.

_____ 23. Mixing methods can be easily substituted in a recipe.

_____ 24. Gluten development would be encouraged in a yeast bread and discouraged in a quick bread.

_____ 25. When using the paddle attachment on an electric mixer, the machine should be on a high setting.

_____ 26. When making a meringue, you would probably use the whisk attachment to the electric mixer.

_____ 27. Stirring is usually a gentle means of mixing that does not require the use of an electric mixer.

_____ 28. Creaming refers to the method of beating water and flour together.

_____ 29. Folding is often the suggested mixing method when the air bubbles in the batter are the primary leavening agent.

_____ 30. A dough hook simulates the creaming process for yeast doughs.

Multiple Choice

Identify the choice (a, b, or c) that most accurately answers the question or completes the statement.

31. The physical act of combining ingredients in a bowl is
     a. mixing
     b. leavening
     c. kneading

32. Overmixing can develop
     a. streaks in the final product
     b. a longer baking time
     c. gluten

33. Undermixing can cause unwanted affects in the final product such as
     a. gluten
     b. too much volume
     c. a lumpy texture

34. Butter, lard, margarine, vegetable shortening, and oils are in a group of ingredients called
     a. starches
     b. fats
     c. carbohydrates

35. An important tool when making pie crust dough is a
     a. pastry blender
     b. pastry bag
     c. dough hook

Matching

Match each term in column 1 with its definition from column 2.

_____ 36. paddle

_____ 37. dough hook

_____ 38. whip

_____ 39. rubber spatula

_____ 40. pastry blender

_____ 41. stirring

_____ 42. folding

_____ 43. mixing

_____ 44. fats

_____ 45. gluten

a. used to simulate kneading process

b. used to incorporate air into the ingredient

c. used primarily to cream and blend

d. tool used to gently mix airy ingredients

e. used to cut fat into flour

f. gives yeast bread its texture

g. the method for incorporating beaten egg whites

h. lard, butter, and oils

i. method that uses a spoon, spatula, or fork

j. the physical act of combining ingredients in a bowl

CHAPTER 5

Thickeners and Stabilizers

After reading this chapter and successfully answering the following questions, you should be able to:

- Define starch.

- Explain the role starch plays as a thickener in desserts.

- Describe the three stages of gelatinization.

- List the factors that affect gelatinization.

- Describe the different types of starch and their characteristics.

- Understand how to choose which starch is best.

- Define gelatin.

- Explain the role gelatin plays as a stabilizer in desserts.

- List the three steps in preparing gelatin.

- Recognize the factors and ingredients that interfere with gelatin.

- Demonstrate how to use starches and gelatin properly by preparing the recipes at the end of this chapter.

A Brief Outline of the Chapter

    I.   Starches defined

   II.   The three stages of gelatinization

  III. Several factors that affect gelatinization

IV. Interesting facts about starches
 V. Gelatin defined
    A. How gelatin forms a gel
    B. How to use gelatin
    C. Other gelatin-like thickeners and stabilizers
    D. Factors and ingredients that interfere with gelatinization

Key Terms

Explain these key terms.

1. alpha-amylase

2. blooming

3. carrageenan

4. gelatin

5. gelatinization

6. gelation

7. gum Arabic

8. gum tragacanth

9. gums

10. pectin

11. starches

Fill in the Blanks

Use the most accurate word or phrase to complete each sentence.

12. The main chemical structure of starch is in two forms. One form is amylose, which has a straighter chemical structure, and _____, which has a branched chemical structure.

13. The varying amounts of amylose compared to amylopectin affect the _____ power of a particular starch.

14. Starches added to processed foods, _____, are a combination of grain and root starches that have chemicals added.

15. The three stages of gelatinization are heating the starch, absorption of liquid, and _____.

16. Overstirring can affect gelatinization by making the liquid _____ because the starch granules break down.

17. Wheat flours may not be as good a thickener as cornstarch because of the _____ content.

18. Gelatin is basically a _____ made from the connective tissue and bones of animals.

19. Gelatin that is used for desserts comes in two forms, powdered granules and _____.

20. The final step in using gelatin for a dessert is called _____ where the mixture firms to a solid like gel.

True or False

Identify each statement as true (T) or false (F).

_____ 21. The amount of liquid used to moisten gelatin is not important.

_____ 22. Vigorous stirring will cause the starch granules to break down.

_____ 23. Gums are used as both thickeners ands stabilizers.

_____ 24. Starches get thicker and reach maximum thickness at 400°F (204°C).

_____ 25. Sugar aids in the process of gelatinization.

_____ 26. Starches have a long shelf life but may lose strength after two years.

_____ 27. Gum tragacanth is derived from animal protein.

_____ 28. The role of a starch in pastry cream is not only to thicken it but to keep the eggs from curdling.

_____ 29. Pectin is found in animal connective tissue and bones.

_____ 30. Pregelatinized starch can withstand freezing and the addition of acidic ingredients without breaking down.

Multiple Choice

Identify the choice (a, b, or c) that most accurately answers the question or completes the statement.

31.   Starches can be derived from all of the following except
   a.   animal connective tissue
   b.   grains
   c.   plant roots

32.   The most common starch used in baking is
   a.   wheat flour
   b.   cornstarch
   c.   arrowroot

33.   Alpha-amylase is an enzyme that affects gelatinization and is found in
   a.   animal connective tissue
   b.   raw eggs
   c.   starches

34.   Although starches need to come to a boil to reach their full thickening potential, too high a temperature could
   a.   break and thin the mixture
   b.   gel the mixture
   c.   reduce the mixture

35.   Which of the following has the least thickening power?
   a.   wheat flour
   b.   cornstarch
   c.   arrowroot

Matching

Match each term in column 1 with its definition from column 2.

_____ 36. gelation

_____ 37. gum Arabic

_____ 38. gum tragacanth

_____ 39. pectin

_____ 40. tapioca

_____ 41. amylose

_____ 42. alpha-amylase

_____ 43. gelatin

_____ 44. gelatinization

_____ 45. blooming

a. process by which starch thickens a liquid

b. a protein

c. an enzyme

d. one of two forms in the structure of starch

e. derived from a plant root

f. found in unripened fruit

g. the firming of gelatin

h. derived from an African tree sap

i. the softening stage

j. derived from a Middle Eastern shrub

CHAPTER 6

Eggs as Thickeners

After reading this chapter and successfully answering the following questions, you should be able to:

- Describe the composition of an egg.

- Understand the importance of proper sanitation when using eggs.

- Define stirred and baked custards.

- Explain how egg proteins become set, coagulate, and thicken custards.

- Describe ways to prevent curdling of custards.

- Demonstrate the steps to prepare a stirred custard.

- Demonstrate the steps to prepare a baked custard.

- Understand how alpha-amylase can thin custards.

- Understand casein and how to avoid its formulation.

- Demonstrate how to properly use eggs as thickeners by preparing the recipes at the end of this chapter.

A Brief Outline of the Chapter

VII.   Coagulation of proteins

VIII. Ways to prevent curdling of custards

IX.   Steps to prepare a stirred custard

X.   Steps to prepare a baked custard

XI.   Problems to avoid with custards

Key Terms

Explain these key terms.

1. air cell

2. alpha-amylase

3. amino acids

4. bain marie

5. baked custard

6. carryover cooking

7. casein

8. coagulation

9. curdle

10. crème anglaise

11. denatured

12. enzymes

13. pasteurization

14. pastry cream

15. porous

16. proteins

17. scalding

18. stirred custard

19. tempering

20. thickeners

Fill in the Blanks

Use the most accurate word or phrase to complete each sentence.

21. A large egg weighs approximately _____.

22. Eggs will keep for approximately _____ if stored at a minimum temperature of 32°F (2°C).

23. The small ropelike material that anchors the yolk is called the

_____.

24. As an egg ages, the _____ at the large end gets larger.

25. Egg sizes include peewee, small, medium, large, extra large, and _____; although the size called for in most recipes is _____.

26. When the cholesterol and fat of the eggs in a recipe must be kept at a minimum, _____ might be used, but read the label carefully as other ingredients may have been added to the product.

27. When eggs are used to help ingredients become less fluid and more dense, they are being used as _____.

28. Any liquid that is thickened by the coagulation of egg proteins is defined as a _____.

29. _____ is the term for any stirred custard that contains a starch.

30. The two categories of custards are _____ and _____.

True or False

Identify each statement as true (T) or false (F).

_____ 31. To determine how fresh an egg is, you must fry it and check how rubbery it is.

_____ 32. Fresh eggs have a yolk that is generally taller than older eggs.

_____ 33. Eggs should be stored away from strong, odorous foods because eggs can absorb the odor.

_____ 34. Raw egg whites can be kept in an airtight container for up to four days.

_____ 35. The highest-grade egg is labeled Grade A.

_____ 36. Some commercially available frozen egg whites contain small amounts of sugar to lower the freezing point.

_____ 37. A crème anglaise is a vanilla sauce.

_____ 38. Dried eggs do not need to be refrigerated.

_____ 39. Frozen whole eggs may contain citric acid to prevent discoloration.

_____ 40. Eggs naturally contain all the essential amino acids that the human body needs for protein formation.

Multiple Choice

Identify the choice (a, b, or c) that most accurately answers the question or completes the statement.

41. The two main roles of eggs in baking are
    a. as a thickener and a leavening agent
    b. as a sweetener and a leavening agent
    c. as a starch and a thickener

42. The weight of an egg is generally
    a. 4 ounces
    b. 2 ounces
    c. ½ ounce

43. A case of eggs contains
    a. 360 eggs
    b. 12 eggs
    c. 12 dozen

44. When an egg is fresh
    a. the yolk is flat and the white is thin
    b. the yolk is thin and the white is firm
    c. the yolk stands high and the white is firm

45. To prevent curdling, most custard sauces should not exceed
    a. 180°F to 185°F
    b. 212°F
    c. 98°F

Matching

Match each term in column 1 with its definition from column 2.

_____ 46. tempering

_____ 47. curdling

_____ 48. bain marie

_____ 49. casein

_____ 50. alpha-amylase

_____ 51. crème anglaise

_____ 52. chalazae

_____ 53. egg yolk

_____ 54. egg white

_____ 55. egg shell

a. a porous covering

b. a stirred custard

c. found in dairy products

d. a warm water bath

e. anchors the yolk in an egg

f. an enzyme in raw eggs

g. an effect of overheating

h. also called albumin

i. contains the fat in an egg

j. a process to gradually increase the temperature of an egg mixture

CHAPTER 7

Eggs as Leaveners and Meringues

After reading this chapter and successfully answering the following questions, you should be able to:

Define an egg foam.

Identify the three types of meringues.

List the tips to ensure a successful meringue.

Describe and recognize the stages of egg foams as they reach soft peaks and stiff peaks.

Define a soufflé.

Define a marshmallow.

Demonstrate the role of eggs in meringues and as leavening agents by preparing the recipes in this chapter.

A Brief Outline of the Chapter

I. Egg foams defined

II. Three types of meringues

    A. French meringues

    B. Swiss meringues

    C. Italian meringues

III. Helpful hints to get the fluffiest meringues

   IV. Using a stainless steel versus a copper bowl to prepare meringues

   V. Separating eggs

   VI. One method to separate eggs

  VII. Substituting pasteurized egg whites

 VIII. How to prepare a safe meringue using fresh shell eggs

   IX. The trouble with underbeaten and overbeaten egg whites

   X. Defining key terms

   XI. Incorporating egg foams into other ingredients

  XII. Defining soufflés

Key Terms

Explain these key terms.

1. cream of tartar

2. egg-foams

3. French meringue

4. hygroscopic

5. Italian meringue

6. marshmallow

7. meringues

8. pasteurization

9. *Salmonella*

10. soft peaks

11. soufflé

12. stiff peaks

13. Swiss meringue

Fill in the Blanks

Use the most accurate word or phrase to complete each sentence.

14. When eggs are beaten, _____ becomes trapped in the _____.

15. Beating eggs _____ the protein by breaking them into fragments.

16. Eggs beaten with sugar added are called _____.

17. _____ peaks tend not to hold their shape and fall over.

18. There are _____ types of meringues used in the bakeshop.

19. The air trapped by whisking egg whites _____ when heated.

20. A _____ is an ingredient within a recipe that helps the final product rise.

21. Eggs with air beaten into them are _____ and are used in many baked goods and desserts.

22. The boiling sugar syrup for a(n) _____ meringue is heated to 240°F to 250°F.

True or False

Identify each statement as true (T) or false (F).

_____ 23. Cream of tartar gives additional flavor to a meringue.

_____ 24. The bowl of the mixture should be perfectly clean of all grease and oils when making meringue.

_____ 25. When more sugar is added to the egg whites, the meringue is firmer.

_____ 26. French meringue is also called common meringue.

_____ 27. It is completely safe to eat common meringue uncooked.

_____ 28. Heat will also denature the egg protein.

_____ 29. Cold eggs make a fluffier meringue than room temperature eggs.

_____ 30. A Swiss meringue would be a good choice for buttercream frosting.

_____ 31. Egg yolks do not make a good leavener.

_____ 32. Egg whites that are underbeaten are likely to collapse.

Multiple Choice

Identify the choice (a, b, or c) that most accurately answers the question or completes the statement.

33. To make an Italian meringue, the sugar should be boiled to at least
    a. 165°F (74°C)
    b. 190°F (88°C)
    c. 240°F (115°C)

34. Egg whites whip better if they are
    a. at room temperature
    b. soft boiled
    c. soaked in hot water for two minutes

35. *Salmonella* is
    a. a bacteria
    b. a neurotoxin
    c. a virus

36. Underbeaten egg whites allow
    a. liquid to leak out of the meringue
    b. for the addition of more sugar
    c. the egg to make a fluffier soufflé

37. When sugar absorbs water, it is said to be
    a. microscopic
    b. thermal
    c. hygroscopic

38. To incorporate another ingredient into meringue, you should
    a. whisk it in at high speed to make more egg foam
    b. whisk it in at low speed so that the meringue does not break
    c. gently fold it in with a rubber spatula

39. Overbeaten egg whites may become
    a. tight and dry
    b. perfect for soufflés
    c. make no difference in the finished product

40. *Salmonella* is destroyed at
    a. 98.6°F (37°C)
    b. 212°F (100°C)
    c. 160°F (71.1°C)

41. A bain marie would be used for which of the following meringues?
    a. Swiss meringue
    b. Italian meringue
    c. French meringue

42. When substituting pasteurized egg whites for a recipe that calls for the whites of eight eggs, you would need
    a. 16 fluid ounces of pasteurized egg whites
    b. 4 fluid ounces of pasteurized egg whites
    c. 8 fluid ounces of pasteurized egg whites

Matching

Match each term in column 1 with its definition from column 2.

_____ 43. Italian meringue

_____ 44. French meringue

_____ 45. Swiss meringue

_____ 46. hygroscopic

_____ 47. soufflé

_____ 48. *Salmonella*

_____ 49. room temperature

_____ 50. cream of tartar

_____ 51. 240°F to 250°F

_____ 52. sugar content

a. a property of sugar

b. common meringue

c. usually baked in a ramekin

d. a bacteria

e. prepared over a bain marie

f. needs a boiling sugar syrup

g. determines the firmness of the meringue

h. assures a fluffier meringue

i. powdered potassium hydrogen tartrate

j. temperature of the syrup in an Italian meringue

CHAPTER 8

Working with Yeast in Straight Doughs

After reading this chapter and successfully answering the following questions, you should be able to:

- Describe the role carbon dioxide plays in the leavening of yeast breads.

- Explain what yeast is.

- List the different types of yeast.

- Define gluten.

- Define gluten's role in baking.

- List ways to control gluten.

- Demonstrate the 12 steps of yeast dough production.

- Show how to best handle a yeast dough when shaping.

- Demonstrate how to work with yeast in straight doughs by preparing the recipes in this chapter.

A Brief Outline of the Chapter

I. Defining yeast

   A. Fresh or compressed yeast

   B. Active dry yeast

   C. Instant active dry yeast

   D. Osmotolerant instant active dry yeast

II. Yeast and fermentation

III. The correct temperature to ferment yeast doughs

IV. How gluten traps carbon dioxide gas

V. Developing flavor in yeast breads

VI. How the amount of water in a yeast dough can affect hole structure

VII. Ingredients that negatively affect yeast breads

VIII. The 12 steps of yeast dough production

IX. Two basic types of yeast doughs

   A. Lean doughs

   B. Rich doughs

Key Terms

Explain these key terms.

1. active dry yeast

2. autolyse

3. baking

4. couche

5. degassing

6. fermentation

7. fresh or compressed yeast

8. friction factor

9. gelatinization of starches

10. glutathione

11. gluten

12. instant active dry yeast

13. kneading

14. lean dough

15. leavener

16. Maillard reaction

17. mixing

18. modified straight dough method

19. osmosis

20. osmotolerant instant active dry yeast

21. oven spring

22. proofing

23. protease

24. resting or benching

25. rich dough

26. scaling

27. scoring or slashing

28. shaping

29. sponge method

30. straight doughs

31. straight dough method

32. wash

33. yeast

Fill in the Blanks

Use the most accurate word or phrase to complete each sentence.

34. Yeast helps doughs and batters rise through the production of
_____.

35. _____ dough ingredients are mixed together in one bowl.

36. A one celled fungus that helps breads rise is _____.

37. Yeast requires _____, _____, and _____ to grow.

38. Active dry yeast has a shelf life of up to _____.

39. _____, or rehydrating, is a good way to tell whether dry yeast is still alive.

40. The optimum time to proof yeast is _____.

─────

True or False

Identify each statement as true (T) or false (F).

_____ 41. If instant active dry yeast is substituted for active dry yeast, the fermentation time will remain about the same.

_____ 42. Instant active dry yeast does not require proofing.

_____ 43. Carbon monoxide is the gas given off by yeast.

_____ 44. The energy created when mixing dough is called friction.

_____ 45. Room temperature has no bearing on the friction factor.

_____ 46. Wheat flour contains two protiens, glutenin and gliadin.

_____ 47. All flours contain gluten-forming proteins.

_____ 48. Fresh compressed yeast can be killed with cold water.

_____ 49. For a rich, sweet dough, an osmotolerant instant active dry yeast would be used.

_____ 50. Yeast doughs will not ferment at room temperatures.

Multiple Choice

Identify the choice (a, b, or c) that most accurately answers the question or completes the statement.

51. The following are by-products of fermentation except
    a. water
    b. carbon dioxide
    c. alcohol

52. The difference between the temperature of the dough before mixing and after mixing is
    a. the Maillard process
    b. the friction factor
    c. the osmotic effect

53. Gluten cannot develop unless
    a. salt is added to the flour
    b. sugar is added to the flour
    c. water is added to the flour

54. A lower protein flour combined with more water
    a. makes the bread bake faster
    b. makes the hole structure larger
    c. makes the gluten stronger

55. Using old, expired yeast
    a. causes a weaker gluten structure for bigger holes
    b. makes a more flavorful bread
    c. makes a dense, flavorless bread

Matching

Match each term in column 1 with its definition from column 2.

_____ 56. degassing

_____ 57. oven spring

_____ 58. Maillard reaction

_____ 59. wash

_____ 60. salt

_____ 61. proofing

_____ 62. glutathione

_____ 63. gluten

_____ 64. a fungus

_____ 65. straight dough

a. the rapid rise of dough when it is put in the oven

b. caramelization of sugars that causes browning

c. brushing the dough with a liquid

d. punching the dough down

e. uses one bowl

f. yeast

g   an amino acid

h. strengthens gluten

i. a test for yeast

j. a network of fibers

CHAPTER 9

Preferments

After reading this chapter and successfully answering the following questions, you should be able to:

- Define a preferment.

- List the benefits of using a preferment.

- Understand the difference between a straight dough and a dough that uses a sponge or starter.

- Recall the differences between the various types of sponges and sourdoughs.

- Work with preferments and prepare the recipes at the end of this chapter.

A Brief Outline of the Chapter

Key Terms

Explain these key terms.

1. *Acetobacillus*

2. artisan breads

3. biga

4. chef

5. friction

6. *Lactobacillus*

7. levain

8. levain-levure

9. natural starters

10. poolish

11. preferments

12. retarding

13. sourdoughs

14. sourdough cultures

Fill in the Blanks

Use the most accurate word or phrase to complete each sentence.

15. Soaking old bread in water, squeezing it dry, and allowing it to ferment is called _____.

16. Sponges are made with _____ yeast.

17. A typical biga contains _____ water.

18. The _____ a poolish is expected to ferment, the _____ yeast is used.

19. Pâté fermentée is simply _____ that has been saved from a previous batch of bread.

20. Two species of bacteria that exist in sourdough are *lactobacillus* and _____.

21. The beginning stages of a sourdough starter is referred to as a _____ or seed culture.

22. In a healthy natural starter, yeast and _____ live in a symbiotic relationship.

23. The yeast in a natural starter gives off alcohol and _____ _____.

24. Straight doughs are mixed _____.

25. Artisan bread makers are associated with _____ because of the flavor they impart and their leavening ability.

True or False

Identify each statement as true (T) or false (F).

_____ 26. Natural and commercial yeasts thrive in an acidic environment.

_____ 27. Starters using commercial yeasts tend to be more reliable than natural yeasts.

_____ 28. Yeast lies dormant between 34° and 40°F (1° and 4°C) without being harmed.

_____ 29. Bacteria are another factor in the final taste of sourdough breads.

_____ 30. Wild yeast starters can be fermented longer than commercial yeast at cooler temperatures.

_____ 31. A thinner sourdough batter ferments at a faster rate than does a thicker batter.

_____ 32. Both sponges and sourdoughs are completely used up in a recipe, so both must be started anew for each batch of bread.

_____ 33. Biga is an Italian word for a type of sourdough starter. ·

_____ 34. A poolish is typically prepared using a flour to water ratio of 4 to 1.

_____ 35. The presence of *acetobacillus* in sourdough ruins the fermentation process.

Multiple Choice

Identify the choice (a, b, or c) that most accurately answers the question or completes the statement.

36. A type of sponge that is simply a piece of dough that has been saved from a previous batch of bread and added to a new batch is called
    a. pâté fermentée
    b. biga
    c. levain-levure

37. In France the sourdough starter would be
    a. a biga
    b. a levain
    c. a chef

38. Barm, desem, or a mother are all terms for
    a. a sponge
    b. yeast
    c. a sourdough starter

39. The sound that a healthy starter should make when stirred is similar to
    a. a soft swoosh
    b. bubble paper being popped
    c. a door slamming

40. The following are characteristics of sponges except
    a. they tend to be short-lived
    b. they tend to be long-lived
    c. they ferment from 30 minutes to several hours

Matching

Match each term in column 1 with its definition from column 2.

_____ 41. chef	a. has two categories: sponges and sourdoughs
_____ 42. biga	
_____ 43. poolish	b. prepared with equal parts of flour and water
_____ 44. straight dough	
_____ 45. artisan breads	c. simple yeast dough
_____ 46. altus brat	d. an Italian preferment
_____ 47. *Lactobacillus*	e. first stage of a sourdough starter
_____ 48. acids	f. usually uses traditional methods
_____ 49. retarding	g. a German sponge
_____ 50. preferment	h. a bacteria
	i. can add flavor
	j. slowing down fermentation

CHAPTER 10

Laminated Doughs

After reading this chapter and successfully answering the following questions, you should be able to:

- Define a laminated dough.

- List three different types of laminated doughs.

- Demonstrate the three-fold and four-fold folding techniques.

- Work with laminated doughs and prepare the recipes at the end of this chapter.

A Brief Outline of the Chapter

Key Terms

Explain these key terms.

1. base dough

2. four-fold or bookfold

3. détrempe

4. laminated dough

5. plasticity

6. three-fold or letterfold

7. turn

Fill in the Blanks

Use the most accurate word or phrase to complete each sentence.

8. Puff pastry base dough is a base dough that contains no _____ or yeast.

9. The letterfold method entails rolling dough into a _____ and folding it like a letter.

10. The three steps in making a laminate dough are preparing the base dough, _____, and making a series of folds to produce thin layers.

11. After making each turn, the dough must be allowed to _____ for _____ to allow the gluten to relax.

12. Laminated doughs can be used for either _____ or _____ baked goods.

13. Danish and croissant doughs are turned _____ times.

14. Puff pastry dough is turned _____ times.

15. The greater the amount of _____, the greater the _____ in pastry.

16. The _____ enclosed will affect the flavor and flakiness of the finished product.

17. Solid fats other than butter are used for three reasons: they _____ _____, have a higher melting point, and do not get as hard.

18. Fats with the highest degree of plasticity are called _____.

19. The ideal temperature of the dough fat is _____.

True or False

Identify each statement as true (T) or false (F).

_____ 20. The type of fat used in making laminated doughs is not important.

_____ 21. One good reason to chill the fat is so that it does not melt while making the other folds.

_____ 22. When making laminated dough, the fat should be colder than the base dough.

_____ 23. To make the dough easier to work, some bakers add lemon juice or vinegar.

_____ 24. The sequence of rolling out a laminated dough and folding it is called a turn.

_____ 25. Danish and croissant doughs contain no yeast.

_____ 26. A four-fold turn is also known as a 4 X 4.

_____ 27. It is necessary to chill the laminated dough only to let the gluten relax.

_____ 28. Laminated doughs should be baked at 400°F (205°C) to allow for the steam to help leaven the final product.

_____ 29. A laminated dough can be made with bread flour, all-purpose flour, or a combination of both.

Multiple Choice

Identify the choice (a, b, or c) that most accurately answers the question or completes the statement.

30. The rising power of Danish and croissant doughs is due to
    a. the sugar
    b. the eggs
    c. the yeast

31. Proper puff pastry dough rises more than Danish or croissant dough
    because
    a. there is no sugar
    b. the yeast
    c. there are more layers

32. If the fat becomes exposed when rolling laminated dough,
    a. coat it with a little flour and continue folding
    b. cut the excess off and continue folding
    c. throw the batch out and start over

33. Always fold the exposed ends of the lamination toward the center so
    that
    a. you can keep track of how many folds you have made
    b. the fat is better contained
    c. to keep the gluten strands oriented in one direction

34. The fat chosen for laminated dough is important because
    a. it affects the flavor of the finished product
    b. it affects the flavor and flakiness of the finished product
    c. it is not important to the final product

35. Lemon juice can be added to the base dough to
    a. add flavor
    b. relax the gluten
    c. add moisture and flavor

36. The number of turns a dough has received should be recorded so that
    a. you know whether you are making a puff pastry or Danish dough
    b. you can tell how much more work is needed to finish the dough
    c. you do not overwork the dough and let the fat get too warm

37. Butter contributes the best taste in laminated doughs because it
    a. has a higher melting point
    b. does not leave a greasy feeling in the mouth
    c. has the greatest plasticity

38. Eggs are added to Danish dough to
    a. enrich the dough
    b. help enclose the fat
    c. add a nice yellow color

Matching

Match each term in column 1 with its definition from column 2.

_____ 39. Danish dough

_____ 40. puff pastry

_____ 41. turning

_____ 42. solid fats

_____ 43. shortening

_____ 44. butter

_____ 45. détrempe

_____ 46. plasticity

_____ 47. letterfold

_____ 48. laminated dough

_____ 49. four-fold

a. laminated dough that uses no sugar, eggs, or yeast

b. a laminated dough that uses eggs, sugar, and yeast

c. butter, vegetable shortening, and margarine

d. the process of layering base dough and fat, rolling it out, and folding i

e. fat of choice for quality laminated doughs

f. another name for base dough

g. ability of fat to hold its shape

h. bookfold turn

i. three-fold turn

j. rich doughs with or without yeast

k. fat high in plasticity

CHAPTER 11

Working with Fats in Pies and Tarts

After reading this chapter and successfully answering the following questions, you should be able to:

- Explain the difference between the three types of pastry crusts: pâte brisée, pâte sucrée, and pâte sablée.

- Explain the role fat plays in making pies and tarts.

- List the six steps to reduce gluten formation.

- Demonstrate how to work with fats in pies and tarts by preparing the recipes in this chapter.

A Brief Outline of the Chapter

I. The difference between tenderness and flakiness in a pastry crust

II. Choosing the right fat

III. Three different types of pastry doughs

    A. Pâte brisée

    B. Pâte sucrée

    C. Pâte sablée

IV. Six ways to ensure a tender, flaky pastry crust

V. Blind baking

VI. Helpful tips to roll out a pastry crust

VII. Preventing a soggy bottom crust

Key Terms

Explain these key terms.

1. blind baking

2. docking

3. galette

4. pâte brisée

5. pâte sablée

6. pâte sucrée

7. stippling

Fill in the Blanks

Use the most accurate word or phrase to complete each sentence.

8. The three types of pastry crusts are _____,
    pâte sucrée, and pâte sablée.

9. A pie without a top crust is called a _____.

10. Tenderness results when the fat shortens the strands of _____,
    preventing them from joining; this is why the fat is referred to as
    _____.

11. Flakiness results when the pieces of fat act as _____ within the
    dough.

12. The two types of pâte brisée are _____, which is best for the
    top crust of the pie, and _____, which is best for the bottom
    crust.

13. To ensure a tender and flaky crust, the fat should be _____.

14. The fat of choice for pie or tart crusts, either by itself or in combinations
    with other fats, is _____.

15. An _____ such as orange or lemon juice is added to the dough
    to prevent gluten formation.

16. Overmixing the dough results in the formation of _____.

17. Baking a pie shell before filling it is called _____.

18. The most common mistake in making pastry dough is to add too much
    _____.

19. When blind baking, stippling or docking the dough helps to prevent
    _____.

20. If the dough becomes sticky during rolling, it can be _____
    for 20 to 30 minutes.

21. Stretching the dough as you fit it into the pan can cause _____
    during baking.

True or False

Identify each statement as true (T) or false (F).

_____ 22. Flakiness and tenderness mean the same thing.

_____ 23. Solid fat is ideal for preparing crusts for pies and tarts.

_____ 24. The flaky type of pâte brisée is perfect for a top crust.

_____ 25. An acid, such as lemon juice or vinegar, is used to ensure a tender, flaky crust.

_____ 26. A high-protein flour is best for making pies and tarts.

_____ 27. The more the dough is kneaded, the flakier and more tender the pie or tart.

_____ 28. Cheddar cheese may be used as a substitute for the fat in a savory pastry dough recipe.

_____ 29. Briefly chilling or freezing a raw pie will reduce shrinkage during baking.

Multiple Choice

Identify the choice (a, b, or c) that most accurately answers the question or completes the statement.

30. The two sweet doughs are
    a. pâte brisée and pâte sucrée
    b. pâte sucrée and pâte sablée
    c. pâte sablée and pâte brisée

31. The fat in the mealy type of pâte brisée is different from the flaky in that
    a. it is almost completely absorbed into the flour
    b. it remains in visible hunks
    c. it allows the flour to absorb moisture

32. Rotating the dough while rolling it out is good because
    a. it prevents the gluten from forming
    b. it keeps the dough from sticking and ensures an even thickness
    c. it helps tenderize the dough

33. One way to prevent the bottom crust from getting soggy during baking
    is to
    a. make sure it is completely baked before removing it from the oven
    b. bake it on the topmost rack in the oven
    c. use the flaky type of pâte brisée

34. To help in rolling out sticky dough, you can
    a. place it on a warm surface
    b. place it between two sheets of plastic wrap
    c. place it on a greased surface

35. The water added in making the dough should be
    a. tepid
    b. hot
    c. ice cold

36. Oil is not used as a fat in making pastry crusts because
    a. it creates a crust that crumbles instead of flakes
    b. it is too expensive
    c. it reacts negatively with the other ingredients

37. Allowing the dough to rest relaxes the gluten and makes
    a. baking time shorter
    b. rolling the dough easier
    c. baking time longer

38. Lard produces a flaky crust but is not often used now because
    a. it is too expensive
    b. it has too low a degree of plasticity
    c. of health concerns over its high saturated fat content

Matching

Match each term in column 1 with its definition from column 2.

_____ 39. pâte sucrée

_____ 40. flaky

_____ 41. well-floured surface

_____ 42. mealy pie dough

_____ 43. stippling

_____ 44. melted chocolate

_____ 45. blind baking

a. needed to properly roll out dough

b. used to prevent puffing

c. possible coating to prevent a soggy bottom

d. a sweet dough

e. characteristic of a good pastry crust

f. fat almost completely absorbed by the flour

g. baking before filling

CHAPTER 12

Using Chemical and Steam Leaveners

After reading this chapter and successfully answering the following questions, you should be able to:

■ Describe what a chemical leavener is.

■ Explain the role leaveners play in baked goods.

■ Differentiate between baking powder, baking soda, and ammonium carbonate.

■ Describe the roles of air and steam in the leavening of baked goods.

■ Define a quick bread.

■ Describe the role carbon dioxide plays in the leavening of quick breads, cakes, and cookies.

■ Demonstrate how to work with chemical and steam leavening agents by preparing the recipes at the end of this chapter.

A Brief Outline of the Chapter

I. Chemical leaveners

II. The important role of air in leavening

III. Steam leavening

IV. Two methods to bake éclair paste

Key Terms

Explain these key terms.

1. ammonium carbonate

2. baking powder

3. baking soda

4. chemical leaveners

5. double-acting baking powder

6. éclair paste or choux paste

7. neutralization reaction

8. pâte à choux

9. popovers

10. quick breads

11. single-acting baking powder

12. steam

Fill in the Blanks

Use the most accurate word or phrase to complete each sentence.

13. Quick breads may be leavened with carbon dioxide gas from
    _____ leaveners.

14. Three common chemical leaveners include ammonium carbonate, baking
    powder, and _____.

15. The two types of baking powder are single-acting, or fast-acting, and
    _____.

16. The exclusive leavening agent in popovers is _____.

17. When baking soda comes in contact with moisture and an acid, a
    _____ takes place.

18. The pH scale is used to measure _____ and _____.

19. Water has a pH of 7 and is considered neutral. _____ have a
    pH of less than 7, and _____ have a pH of higher than 7.

20. Baking powder is a mixture of baking soda and one or more
    _____.

21. Mixing in any form incorporates _____ into the substance.

22. Éclair paste that is formed into small rounds or puffs are referred to as
    _____.

23. In making éclair paste, _____ are added for
    leavening and to provide structure.

True or False

Identify each statement as true (T) or false (F).

_____ 24. Quick breads refer to baked goods that use only yeast as a leavening agent.

_____ 25. Chemical leaveners react with moisture and heat to produce carbon monoxide gas.

_____ 26. A substance with a pH of 8 is a base.

_____ 27. Neutralization reactions occur when equal amounts of acids and bases are combined.

_____ 28. Sodium bicarbonate and ammonium carbonate are both acids.

_____ 29. Double-acting baking powder works faster than single-acting baking powder.

_____ 30. You can never overmix a quick bread.

_____ 31. The two methods for baking éclair paste differ on the temperature of the oven.

_____ 32. To neutralize a batter that contains a large amount of buttermilk, you would add baking powder.

_____ 33. Baking soda retains its strength indefinitely.

_____ 34. Ammonium carbonate is best used for items that will be baked at a high temperature until dry and crisp.

Multiple Choice

Identify the choice (a, b, or c) that most accurately answers the question or completes the statement.

35. Scones, muffins, and shortcakes are examples of
    a. yeast breads
    b. quick breads
    c. laminated dough

36. Three chemical leaveners are
    a. baking soda, baking powder, and yeast
    b. bicarbonate of soda, ammonium carbonate, and sugar
    c. baking powder, baking soda, ammonium carbonate

37. Yogurt and buttermilk are examples of
    a. acidic ingredients
    b. bases
    c. chemical leaveners

38. Sodium bicarbonate is another name for
    a. baking soda
    b. ammonium carbonate
    c. carbon monoxide

39. The two types of baking powder are
    a. double-acting and triple-acting
    b. single-acting and retroacting
    c. single-acting and double-acting

40. One major difference between baking soda and baking powder is that baking powder
    a. contains sugar
    b. contains one or more acids
    c. contains ammonium carbonate

41. In puff pastry, the exclusive leavening agent is
    a. yeast
    b. baking soda
    c. steam

42. This chemical leavener should not be used for cakes or muffins:
    a. ammonium carbonate
    b. baking soda
    c. bicarbonate of soda

43. Pâte à choux is the same as
    a. popover batter
    b. éclair paste
    c. laminated dough

44. To make profiteroles, you would use
    a. éclair paste
    b. base dough
    c. popover batter

Matching

Match each term in column 1 with its definition from column 2.

_____ 45. choux paste
_____ 46. base
_____ 47. acid
_____ 48. popovers
_____ 49. water
_____ 50. cream of tartar

a. buttermilk, yogurt, or lemon juice

b. has a pH higher than 7

c. has a pH of 7

d. éclair paste

e. a puffy quick bread

f. fast-acting baking powder ingredient

CHAPTER 13

Quick Bread Mixing Methods

After reading this chapter and successfully answering the following questions, you should be able to:

- Understand the differences between yeast breads and quick breads.

- Understand the three quick bread mixing methods.

- Demonstrate the quick bread mixing methods by preparing the recipes in this chapter.

A Brief Outline of the Chapter

Key Terms

Explain these key terms.

1. biscuit method

2. creaming

3. creaming method

4. cutting

5. muffin method

6. popover

7. tunneling

Fill in the Blanks

Use the most accurate word or phrase to complete each sentence.

8. The three quick bread mixing methods are the biscuit mixing method, the muffin mixing method, and the _____ mixing method.

9. Gluten and thus toughness will result if a quick bread batter is _____.

10. Chemical leaveners work using chemical reactions among the ingredients, whereas _____ is a living organism and uses a fermentation process.

11. In yeast doughs, the rate of fermentation depends on the _____ of the dough.

12. Breads using yeast rise before baking, but quick breads rise _____ baking.

13. The special mixing methods for quick breads are designed to reduce _____ formation.

14. When making scones, the _____ mixing method would be used.

15. When making pancakes, the _____ mixing method would probably be used.

16. The "short" in shortcake comes from the _____ that coats the gluten strands.

17. The fat used in the muffin mixing method is usually _____ in form.

18. In the creaming method, the two ingredients that are creamed are _____ and _____.

19. The technique to combine fat into dry ingredients is _____.

20. When making muffins, the pans should be filled _____ full.

21. One unwanted outcome from overmixing quick breads is _____, which are large holes or cavities throughout the inside of the item.

True or False

Identify each statement as true (T) or false (F).

_____ 22. Quick breads have the same texture as yeast breads.

_____ 23. Quick bread recipes will usually have an instruction to "beat thoroughly with an electric mixer on high speed."

_____ 24. As their name suggests, quick breads are prepared quickly and baked soon after they are mixed.

_____ 25. Yeast breads rise during the baking process and not before.

_____ 26. Gluten development is encouraged in most quick breads.

_____ 27. Blending a sugar and a fat until light and fluffy is called creaming.

_____ 28. Toughness is one possible outcome of overmixing quick breads.

_____ 29. A pastry blender or an electric mixer with a pastry blender attachment would be used in the biscuit mixing method.

_____ 30. The fat used in the biscuit mixing method is usually liquid in form.

_____ 31. Paper liners for muffin pans ensure that the muffins will be larger than they would be without the liners.

Multiple Choice

Identify the choice (a, b, or c) that most accurately answers the question or completes the statement.

32. One difference in texture between yeast breads and quick breads is
    a. yeast breads are chewier
    b. quick breads are chewier
    c. quick breads are dryer

33. The general goal of the three quick bread mixing methods is to
    a. minimize gluten formation
    b. maximize gluten formation
    c. neutralize acids

34. The common leaveners for quick breads are
    a. yeast and eggs
    b. yeast and steam
    c. chemical leaveners and steam

35. The method for making a flaky pie crust is similar to
    a. the creaming mixing method
    b. the biscuit mixing method
    c. the muffin mixing method

36. Having all the liquid ingredients in one bowl and dry ingredients in
    another bowl and then mixing the liquid ingredients into the dry only
    until blended describes
    a. the creaming mixing method
    b. the muffin mixing method
    c. the biscuit method

37. The leavening agent in popovers is
    a. ammonium carbonate
    b. bicarbonate of soda
    c. steam

38. One cause of tunneling in a quick bread is
    a. undermixing
    b. overmixing
    c. underbaking

39. A muffin with a cakelike texture was probably made using
    a. the muffin mixing method
    b. the creaming method
    c. the biscuit mixing method

40. Paper liners for muffins are used commercially to
    a. create a higher and fuller appearance
    b. protect the muffins and prevent them from drying out
    c. lessen baking time

41. Shortcakes generally use the
    a. creaming method
    b. muffin mixing method
    c. biscuit mixing method

Matching

Match each term in column 1 with its definition from column 2.

_____ 42. brown sugar	a.	a chemical leavener
_____ 43. creaming method	b.	a living organism
_____ 44. muffin mixing method	c.	leavener in popovers
_____ 45. biscuit mixing method	d.	an acid ingredient
_____ 46. baking soda	e.	produced by chemical leaveners
_____ 47. yeast	f.	generally uses a liquid fat
_____ 48. carbon dioxide	g.	caused by overmixing
_____ 49. steam	h.	resembles the method to make flaky pie crust
_____ 50. tunneling	i.	creates a cakelike texture
_____ 51. tenderness	j.	result of minimizing gluten formation

CHAPTER 14

Cake Mixing Methods

After reading this chapter and successfully answering the following questions, you should be able to:

- Define the two categories of cakes.

- Describe the three cake mixing methods for cakes high in fat.

- Describe the three cake mixing methods for cakes low in fat.

- Understand the three ways to tell whether a cake is done.

- Prepare recipes using the various mixing methods in this chapter.

A Brief Outline of the Chapter

B. The chiffon method

C. The angel food method

VI. How to tell when a cake is done

Key Terms

Explain these key terms.

1. angel food method

2. cakes high in fat

3. cakes low in fat

4. chiffon method

5. creaming method

6. egg-foam cakes

7. emulsion

8. emulsifying agent

9. high ratio cake

10. immiscible liquids

11. one-stage method

12. sponge method

13. two-stage method

Fill in the Blanks

Use the most accurate word or phrase to complete each sentence.

14. The two categories of cakes are those _____ and those _____ in fat.

15. For cakes high in fat, the three mixing methods are the creaming method, the one-stage method, and the _____.

16. For cakes low in fat, the three mixing methods are the chiffon method, the angel food method, and the _____.

17. A cake that has more sugar than flour is called a _____.

18. In the one-stage method, _____ is the ideal fat to use.

19. The two types of sponge cake include whole egg foams and _____.

20. A _____ is a French sponge cake that has a strong structure.

True or False

Identify each statement as true (T) or false (F).

_____ 21. The creaming method is used for cakes high in fat.

_____ 22. Creaming the sugar and fat for too long can cause a coarse texture in the finished cake.

_____ 23. The two-stage method of mixing is used for cakes low in fat.

_____ 24. Eggs can be used as emulsifying agents.

_____ 25. Egg foams can be made only from egg whites.

_____ 26. Chiffon cakes get their leavening from two sources—a solid fat and water.

_____ 27. Angel food cakes generally use whole eggs.

_____ 28. A cake is done if it springs back when gently pressed.

_____ 29. A thin wooden skewer can be used as a cake tester.

_____ 30. To be called a cake, an item must be baked in 8- or 9-inch round pans.

Multiple Choice

Identify the choice (a, b, or c) that most accurately answers the question or completes the statement.

31. Cakes high in fat use solid or liquid fats to keep
    a. gluten development at a minimum
    b. cooking time to a minimum
    c. costs to a minimum

32. The tenderizer in cakes low in fat is usually
    a. gluten
    b. milk
    c. sugar

33. Egg-foam cakes are different than cakes high in fat in that
    a. they are generally dryer and more flexible
    b. they tend to be moister and denser
    c. they tend to have a coarser crumb

34. The natural tendency for fat and water to separate is called
    a. pH levels
    b. surface tension
    c. emulsification

35. The creaming method is the method of choice for
    a. chiffon cakes
    b. genoise
    c. butter cake

36. In the sponge method, the eggs are warmed to
    a. reduce the cooking time
    b. decrease the volume
    c. increase the volume

37. The advantage of shortening over butter in high fat cakes is that
    a. it has a better flavor
    b. it promotes gluten formation
    c. it already contains evenly dispersed small air bubbles

38. A good example of a pair of immiscible liquids is
    a. eggs and milk
    b. water and oil
    c. milk and water

39. The part of the egg that contains natural emulsifiers is
    a. the yolk
    b. the white
    c. the foam

40. Chiffon cakes get their leavening from these two sources:
    a. air and gluten
    b. sugar and butter
    c. air and baking powder

Matching

Match each term in column 1 with its definition from column 2.

_____ 41. chiffon cake

_____ 42. genoise

_____ 43. egg yolk

_____ 44. butter cake

_____ 45. one-stage method

_____ 46. shortening or butter

_____ 47. a wooden skewer

_____ 48. baking powder

_____ 49. egg foams

_____ 50. angel food cake

a. an example of an emulsifier

b. made using the creaming method

c. usually made with oil

d. may be used to test for doneness in a cake

e. a chemical leavener

f. best fats for the creaming method

g. uses only egg whites

h. a mixing method for cakes high in fat

i. made by beating eggs with sugar

j. a French sponge cake

CHAPTER 15

Frostings

After reading this chapter and successfully answering the following questions, you should be able to:

- Define a frosting.

- State the four reasons to use frostings.

- List the seven categories of frostings.

- Correctly pair cakes with an appropriate frosting.

- Identify the basic tools needed to properly frost a cake.

- Frost a cake.

- Prepare the frosting recipes in this chapter.

A Brief Outline of the Chapter

Key Terms

Explain these key terms.

1. acetate

2. buttercream

3. chocolate confectionery frosting

4. egg-foam or boiled frosting

5. frosting

6. ganache

7. icing comb

8. metal cake rings or torte ring

9. metal false-bottom tart pan bottom

10. offset palette knife

11. palette knife

12. pastry bag

13. pastry brush

14. pastry tips

15. rolled-out frostings

16. royal icing

17. simple or flat icing

18. turntable

Fill in the Blanks

Use the most accurate word or phrase to complete each sentence.

19. The word _____ can be used interchangeably with the word icing.

20. One of the four reasons to use frosting is to _____ the staling process.

21. The seven categories of icings include simple or flat icings, glazes, royal icings, buttercreams, egg-foam or boiled frostings, rich chocolate confectionery frostings, and _____ frostings.

22. A Napoleon is often iced with a _____, which is a special type of _____.

23. A _____ made with sugar syrup, and thinned preserves or jams may be brushed onto a fruit tart.

24. Resembling a fluffy meringue, _____ may be piped into decorations and hardens quickly.

25. The _____ is a type of buttercream frosting that is similar to an Italian meringue.

26. A _____ is a buttercream frosting that begins with a custard thickened with starch.

27. Egg-foam or boiled frostings consist of either _____ or _____ meringues.

28. Rich chocolate confectionery frostings are based on two types of chocolate candy: _____ and _____.

29. The most important factor in the denseness of a ganache frosting is the proportion of _____ to chocolate.

30. Made from finely ground almonds, _____ is one of three types of rolled-out frosting.

True or False

Identify each statement as true (T) or false (F).

_____ 31. Heavier cakes are generally paired with lighter frostings like a simple icing or a glaze, and lighter cakes are frosted with a heavier frosting such as a fudge-style frosting.

_____ 32. One reason to frost a cake may be to increase its eye appeal.

_____ 33. A simple or flat icing must be boiled until it reaches the soft ball stage.

_____ 34. Poured fondant is a sugar syrup that is allowed to crystallize to form a paste.

_____ 35. French buttercream begins with beaten egg yolks.

_____ 36. Fudge is exactly the same as ganache.

_____ 37. Modeling chocolate is made from melted chocolate and corn syrup.

_____ 38. Ganache frosting is delicate and will be ruined if frozen.

_____ 39. A fudge frosting recipe may not contain chocolate.

_____ 40. Truffles use a much higher ratio of cream to chocolate than ganache.

Multiple Choice

Identify the choice (a, b, or c) that most accurately answers the question or completes the statement.

41. The rectangular or triangular piece of metal or plastic that is used to make grooves and ridges in the frosting is called
    a. an icing comb
    b. a palette knife
    c. a pastry bag

42. A thin pourable frosting that can be poured over coffee cakes or scones is
    a. French buttercream
    b. egg-foam
    c. simple icing

43. The four types of buttercream icing are
    a. fudge, simple, marzipan, and ganache
    b. simple, French, Italian, and pastry cream based
    c. egg-foam, simple, Italian, and rolled-out

44. Marzipan, modeling chocolate, and rolled fondant are three types of
    a. rolled-out frosting
    b. buttercream
    c. egg-foam

45. The general ratio of heavy cream to chocolate for ganache as a frosting is
    a. 1 part heavy cream to 2 parts chocolate
    b. 1 part heavy cream to 1 part chocolate
    c. 2 parts heavy cream to 1 part chocolate

46. The only one of the seven types of frosting that is kneaded is
    a. egg-foam
    b. simple or flat icing
    c. rolled-out

47. An offset palette knife differs from a regular palette knife in that
    a. there is a sharper tip on the offset palette knife
    b) there is a serrated edge on the offset pallete knife
    c. there is a bend in the offset palette knife

48. Thin plastic sheets that protect the sides of a cake during frosting are
    a. acetates
    b. bakelites
    c. pastry tips

49. Beating a fudge frosting while it is still hot may cause
    a. the ingredients to separate
    b. a grainy texture
    c. an unpleasant taste

50. A good tool with which to apply a glaze is
    a. a pastry bag
    b. a pastry brush
    c. acetate

Matching

Match each term in column 1 with its definition from column 2.

_____ 51. flat icing
_____ 52. glazes
_____ 53. pastry cream-based buttercream
_____ 54. French buttercream
_____ 55. Italian buttercream
_____ 56. truffles
_____ 57. rolled fondant
_____ 58. marzipan
_____ 59. fudge
_____ 60. modeling chocolate

a. fine ground almonds
b. melted chocolate and corn syrup
c. consists of sugar syrup, jam, or preserves
d. starts with beaten egg yolks
e. ganache
f. similar to meringue
g. starts with sugar syrup boiled to the soft ball stage
h. a cooked sugar mixture that is kneaded
i. thin and pourable
j. starts with a custard that is thickened

CHAPTER 16

Cookies

After reading this chapter and successfully answering the following questions, you should be able to:

- Define a cookie.

- List the eight categories of cookies.

- Explain the three basic cookie mixing methods.

- Manipulate ingredients to attain desirable cookie characteristics.

- Prepare the recipes at the end of the chapter.

A Brief Outline of the Chapter

    I.   Categories of cookies

        A. Drop cookies

        B. Refrigerator cookies

        C. Molded cookies

        D. Bar cookies

        E. Sheet cookies

        F. Rolled cookies

        G. Piped cookies

        H. Wafer cookies

   II.   Three basic cookie mixing methods

        A. Creaming method

        B. One-bowl method

        C. Egg-foam method

III. Understanding the characteristics of cookies and how to manipulate them

IV. Tips for making successful cookies

Key Terms

Explain these key terms.

1. bar cookies

2. cookies

3. creaming method

4. drop cookies

5. egg-foam method

6. molded cookies

7. one-bowl method

8. piped cookies

9. refrigerator cookies

10. rolled cookies

11. sheet cookies

12. wafer cookies

Fill in the Blanks

Use the most accurate word or phrase to complete each sentence.

13. The eight categories of cookies are differentiated on how the _____ is prepared and shaped for baking.

14. The eight categories of cookies include drop cookies, refrigerator cookies, molded cookies, bar cookies, sheet cookies, rolled cookies, _____, and _____ .

15. In preparing drop cookies, the moist dough is literally _____ from the spoon or ice cream scooper onto the sheet pan.

16. Cookie dough that is rolled into a log and then chilled is the definition of a _____.

17. A fudge brownie is a good example of the _____ category of cookie.

18. Cookie dough that is stiff and formed into a long, flattened, rectangular bar and then baked and cut into slices is the _____ category.

19. The batter may be referred to as a tulipe or stencil batter for the _____ category.

20. Forming cookies by pushing the soft dough through a pastry bag is the _____ category.

True or False

Identify each statement as true (T) or false (F).

_____ 21. Sheet cookies are one of those labor-intensive categories of cookies.

_____ 22. Molded cookies are made from a thin, pourable batter.

_____ 23. Drop cookies must be spaced on the baking sheet to allow for spreading.

_____ 24. A cookie cutter is usually used with the rolled cookie category.

_____ 25. Wafer cookies may be so thin as to be translucent.

_____ 26. A tuile is a type of rolled cookie.

_____ 27. In a bar cookie, the dough is cut into shapes prior to baking.

_____ 28. A pastry bag is necessary to form piped cookies.

_____ 29. A wafer cookie may be molded while still hot from the oven.

_____ 30. Molded cookies are usually rolled into balls before baking.

Multiple Choice

Identify the choice (a, b, or c) that most accurately answers the question or completes the statement.

31. The three cookie mixing methods are
    a. the piped method, the rolled method, and the sheet method
    b. the creaming method, the one-bowl method, and the egg-foam method
    c. the wafer method, the egg-foam method, and the two-bowl method

32. All of these factors will increase the spread of a cookie except
    a. using a flour with high protein
    b. increasing the amount of liquid
    c. using a higher amount of baking soda

33. One factor in producing a crisp cookie is
    a. to keep the cookie thin
    b. to keep the granulated sugar to a minimum
    c. to lower the amount of baking soda

34. A cookie that is slightly underbaked will be
    a. crisper
    b. browner
    c. softer

35. Flours that are bleached and more acidic will produce
    a. a crisper cookie
    b. a browner cookie
    c. a paler cookie

36. Due to the coagulation of proteins, the more eggs in the cookie dough
    a. the chewier the cookie
    b. the browner the cookie
    c. the paler the cookie

37. Confectioners' sugar contains
    a. flour
    b. corn starch
    c. baking soda

38. One method to keep a cookie from coloring during baking is
    a. to maintain acidity in the dough
    b. to eliminate acidity in the dough
    c. to increase the baking soda in the dough

39. Overmixing the fat and sugar in the creaming method for cookies could result in
    a. decreased spreading
    b. increased spreading
    c. a paler cookie

40. Hygroscopic sugars are sugars that
    a. cannot absorb moisture from the air
    b. easily absorb moisture from the air
    c. produce crisper cookies

Matching

Match each term in column 1 with its definition from column 2.

_____ 41. drop cookie
_____ 42. refrigerator cookie
_____ 43. piped cookie
_____ 44. molded cookie
_____ 45. sheet cookie
_____ 46. hygroscopic sugars
_____ 47. egg-foam method
_____ 48. cake flour
_____ 49. bread flour
_____ 50. baking powder

a. requires a pastry bag
b. must be chilled or frozen
c. honey and molasses
d. sour cream fudge cookies
e. high in protein
f. low in protein
g. used when an acid is needed
h. starts with beating eggs and sugar
i. truffled peanut butter cookies
j. fudge brownies

CHAPTER 17

Building Blocks with Sugar

After reading this chapter and successfully answering the following questions, you should be able to:

■ Demonstrate the three basic steps to prepare sugar syrups for desserts and confections.

■ Recognize when a sugar syrup is done.

■ Describe the process of crystallization.

■ List three ways to control crystallization.

■ Prepare caramelized sugar using the dry method and the wet method.

■ Prepare the recipes in this chapter.

A Brief Outline of the Chapter

VI.   Where caramel gets its flavor

VII.   Using caution when working with sugar

VIII.   Tips for preparing sugar syrups

IX.   Storage of confections made from caramel

Key Terms

Explain these key terms.

1.   caramelized sugar

2.   cold-water test

3.   crystallization

4.   dry method

5.   interfering agents

6.   inversion

7. invert sugar

8. simple syrup

9. solute

10. solvent

11. sucrose

12. sugar syrup

13. wet method

Fill in the Blanks

Use the most accurate word or phrase to complete each sentence.

14. Crystals of granulated sugar cannot be changed until they are
_____ in water and _____.

15. The three basic steps in preparing sugar syrups for desserts and confections are dissolving the sugar in water to make a sugar syrup, concentrating the sugar syrup, and _____ the sugar syrup.

16. Sugar syrups, one or more sugars dissolved in water, can be called _____.

17. A solution consists of two parts: a _____ and a _____.

18. Sugar syrup is a solution where the sugar is the _____ and the water is the _____.

19. Increasing the _____ of a sugar syrup means that more sugar can be dissolved in it because the _____ keeps the sugar molecules moving fast enough to prevent them from joining together to make crystals.

20. Boiling a sugar syrup accomplishes two things: It allows the sugar to completely dissolve in the water; and it causes some of the water to _____, thus concentrating the solution.

21. Care must be taken during the cooling process of a sugar syrup so that the formation of large _____ can be avoided.

True or False

Identify each statement as true (T) or false (F).

_____ 22. The temperature to which a sugar syrup is cooked must be 400°F.

_____ 23. A candy thermometer or the cold-water test can be used to determine whether the sugar syrup has reached the correct temperature.

_____ 24. To achieve a smooth texture in a confection or dessert, the sugar syrup should be allowed to form large crystals.

_____ 25. One way to control crystallization is to stir the sugar syrup constantly upon heating.

_____ 26. Adding corn syrup to the sugar syrup will prevent crystals from forming.

_____ 27. Adding cream of tartar to a sugar syrup while heating will cause inversion to occur.

_____ 28. You can prevent crystals from sticking to the sides of the pot by oiling the slides.

_____ 29. After a sugar syrup has completed cooking, there is no interfering agent that will work to prevent crystallization.

_____ 30. A good example of an invert sugar is sucrose.

Multiple Choice

Identify the choice (a, b, or c) that most accurately answers the question or completes the statement.

31. To change the crystals of granulated sugar, first it must be dissolved in water, and second
    a. it must be mixed with flour
    b. it must be allowed to settle
    c. it must be brought to a boil

32. The basic ingredients of a simple syrup are
    a. water and brown sugar
    b. water and granulated sugar
    c. water and confectioner's sugar

33. As the water evaporates during boiling, the sugar syrup becomes
    a. more concentrated
    b. more diluted
    c. less dense

34. The darker the caramelized sugar syrup,
    a. the harder it becomes after it has cooled
    b. the softer it becomes after it has cooled
    c. the less concentrated it becomes after it has cooled

35. A cold-water test is used to
    a. test the acidity of a sugar syrup
    b. test the flavor of a sugar syrup
    c. test the temperature or doneness of the sugar syrup

36. When an acid is heated with granulated sugar, it will break down into
    a. lactose and fructose
    b. glucose and fructose
    c. dextrose and maltose

37. The wet method to make caramelized sugar starts with
    a. a sugar syrup boiled to the caramel stage
    b. granulated sugar placed in a sauté pan
    c. egg whites and sugar

38. The dry method to make caramelized sugar starts with
    a. a sugar syrup boiled to the caramel stage
    b. granulated sugar placed in a sauté pan
    c. egg whites and sugar

39. Adding dairy products to caramelized syrups not only helps prevent
    some crystallization but also
    a. decreases the caramel flavor
    b. intensifies the caramel flavor
    c. increases the temperature

40. A safety tip when working with hot sugar syrups is to have
    a. a bowl of ice water nearby
    b. a side towel nearby
    c. a desiccant or drying agent nearby

Matching

Match each term in column 1 with its definition from column 2.

_____ 41. granulated sugar

_____ 42. simple syrup

_____ 43. solvent

_____ 44. solute

_____ 45. an invert sugar

_____ 46. interfering agent

_____ 47. inversion process

_____ 48. Maillard reaction

_____ 49. the wet method

_____ 50. the dry method

a. sugar in a solution of sugar and water

b. water in a solution of sugar and water

c. sugar and water

d. corn syrup

e. occurs when proteins and sugar are heated to a high temperature

f. boiling sugar syrup to the caramel stage

g. starts with sugar and a sauté pan

h. adding acid during preparation of a sugar syrup to prevent crystallization

i. cream of tartar

j. sucrose

CHAPTER 18

Frozen Desserts

After reading this chapter and successfully answering the following questions, you should be able to:

- Explain the difference between churn-frozen and still-frozen desserts.

- Explain the factors that contribute to texture in frozen desserts.

- Describe the roles that sugar and salt play in preparing frozen desserts.

- Demonstrate how to prepare the churn-frozen and still-frozen desserts in this chapter.

A Brief Outline of the Chapter

Key Terms

Explain these key terms.

1. bombe

2. churn-frozen

3. granité (granita)

4. mousse

5. mouthfeel

6. overrun

7. parfait

8. semifreddo

9. sherbet

10. sorbet

11. still-frozen

Fill in the Blanks

Use the most accurate word or phrase to complete each sentence.

12. The two general categories of frozen desserts are churn-frozen desserts
     and _____-frozen desserts.

13. Ice cream is an example of a _____-frozen dessert.

14. When preparing an uncooked custard base for a frozen dessert,
     _____ should be used to prevent *Salmonella*.

15. _____ is a rich Italian ice cream that is denser than American
     ice cream.

16. The _____ in the base of a sorbet prevents it from freezing
     into a block of ice.

17. By law, a sherbet must contain less than _____ milkfat.

18. _____ is the Italian name for a coarse, icy dessert.

19. The three bases for a still-frozen dessert include the custard base, the Italian meringue base, and the _____ base.

20. In a still-frozen dessert, the custard base is often folded into

_____ .

21. A European style parfait uses an _____ base.

True or False

Identify each statement as true (T) or false (F).

_____ 22. By law, ice cream must contain at least 10 percent milkfat.

_____ 23. Frozen yogurt tends to contain a higher percentage of milkfat than ice creams.

_____ 24. Sorbet may be served between courses as a palate cleanser.

_____ 25. Sherbets are generally fruit flavored and have higher milkfat than ice cream.

_____ 26. Still-frozen desserts must be constantly agitated during the freezing process.

_____ 27. A semifreddo is a churn-frozen dessert.

_____ 28. A custard base may be used for churn- and still-frozen desserts.

_____ 29. The smaller the ice crystals in the frozen dessert, the smoother the mouthfeel.

_____ 30. The lower the fat, the creamier the texture of the frozen dessert.

_____ 31. Increasing the volume by incorporating air into the ice cream is called overrun.

Multiple Choice

Identify the choice (a, b, or c) that most accurately answers the question or completes the statement.

32. The ideal time to age a base for a frozen dessert is
    a. 1 hour
    b. 12 to 24 hours
    c. 0, bases should not be aged

33. An example of a good emulsifier is
    a. egg yolks
    b. water
    c. milk

34. The general rule is that the greater the amount of sugar added to a frozen dessert,
    a. the harder the dessert
    b. the grainier the dessert
    c. the softer the dessert

35. The faster the base of a frozen dessert freezes, the smaller the ice crystals, resulting in
    a. a smoother dessert
    b. a grainier dessert
    c. a harder dessert

36. Frozen desserts are best stored at
    a. 40° to 50°F
    b. −10° to 0°F
    c. 20° to 30°F

37. Adding alcohol or liquor to flavor a frozen dessert will
    a. make it easier to freeze
    b. make it more difficult to freeze
    c. not affect the freezing temperature

38. The slowing down of food molecules at lower temperatures means that frozen desserts are
    a. more difficult to flavor than room temperature desserts
    b. less difficult to flavor than room temperature desserts
    c. never as flavorful as room temperature desserts

39. The role of a stabilizer added to ice cream is to
    a. prevent the ice cream from freezing
    b. prevent the ice cream from melting
    c. prevent the ice cream from forming large, grainy ice crystals

40. The percentage of overrun in a premium ice cream is generally
    a. lower than ice creams of lesser quality
    b. higher than ice creams of lesser quality
    c. the same as ice creams of lesser quality

41. By law, the overrun allowed for ice cream is
    a. 50 percent
    b. 200 percent
    c. 100 percent

Matching

Match each term in column 1 with its definition from column 2.

_____ 42. vegetable gum
_____ 43. egg yolks
_____ 44. semifreddo
_____ 45. granitá
_____ 46. sorbet
_____ 47. gelato
_____ 48. overrun
_____ 49. mouthfeel
_____ 50. sherbet
_____ 51. strawberry purée

a. fruit base
b. a rich Italian ice cream
c. increases the volume of ice cream
d. frozen mousse
e. an example of a stabilizer in ice cream
f. an example of an emulsifier in ice cream
g. a grainy, course frozen dessert
h. contains no dairy products
i. describes the texture of a food
j. contain less than 2 percent dairy products

CHAPTER 19

Chocolate

After reading this chapter and successfully answering the following questions, you should be able to:

- Understand the process of chocolate making.

  Know the differences between milk, semisweet, and white chocolates.

  Define how cocoa butter affects the quality of chocolate.

- Demonstrate what tempering is.

- List the difficulties of working with chocolate.

- Demonstrate how to prepare the recipes in this chapter.

A Brief Outline of the Chapter

Key Terms

Explain these key terms.

1. baking chocolate

2. bittersweet chocolate

3. bloom

4. cacao bean

5. chocolate

6. chocolate liquor

7. cocoa butter

8. compound coating

9. couverture

10. Dutch-processed cocoa powder

11. milk chocolate

12. natural cocoa powder

13. nibs

14. seizing

15. semisweet chocolate

16. tempering

17. white chocolate

Fill in the Blanks

Use the most accurate word or phrase to complete each sentence.

18. In the chocolate making process, the _____ are crushed to form a dark liquid called _____.

19. _____, a saturated fat, gives chocolate its velvety texture.

20. High-quality chocolate with at least 32 percent cocoa butter that can be used as a coating is known as _____, whereas the lower-quality chocolate product, also used as a coating, with little or no cocoa butter present is known as _____.

True or False

Identify each statement as true (T) or false (F).

_____ 21. There is a high percentage of alcohol in the chocolate liquor that is made when the nibs of the cocoa beans are crushed.

_____ 22. Dutch processed cocoa must, by law, be made in the Netherlands.

_____ 23. The fat content of cocoa powder may vary but usually has a minimum of 10 percent cocoa butter.

_____ 24. White chocolate contains no chocolate liquor at all.

_____ 25. Just one drop of water can make melted chocolate harden and become discolored.

_____ 26. High-quality chocolates need a temperature of above 115°F to melt.

_____ 27. Chocolate chips can be substituted for high-quality chocolate in any recipe without a problem.

_____ 28. Unsweetened chocolate, baking chocolate, and bittersweet chocolate are all the same.

_____ 29. Milk chocolate must have a minimum of 12 percent milk solids.

_____ 30. Depending on the quality, white chocolate may or may not contain cocoa butter.

Multiple Choice

Identify the choice (a, b, or c) that most accurately answers the question or completes the statement.

31. Placing a cover over a container of warm melted chocolate is not advised because
    a. steam may form and drip into the chocolate
    b. the air cannot get to the chocolate
    c. the chocolate will overheat

32. Tempering is done to chocolate to
    a. eliminate acids
    b. stabilize the crystals in the cocoa butter
    c. eliminate the crystals in the cocoa butter

33. Tempering is a three-stage process of
    a. stirring, heating, stirring
    b. melting, cooling, rewarming
    c. chilling, heating, rechilling

34. The two methods of tempering are the table method and
    a. the Maillard process
    b. the press method
    c. the seeding or injection method

35. The properties that best describe properly tempered chocolate are
    a. a dull but deep color and brittle
    b. shiny and pliable
    c. shiny and breaks with a snap

36. When chocolate goes out of temper, it may develop a bloom, which is
    a. a shiny surface
    b. a whitish-gray spotty coating
    c. an increase in size

37. A sugar bloom may develop on tempered chocolate if
    a. it is left uncovered in the refrigerator
    b. it is kept in an airtight container
    c. the temperature was kept at 70°F

38. The center of a truffle confection is usually made of
    a. milk chocolate
    b. baking chocolate
    c. ganache

39. The two types of truffle centers are
    a. light and dense
    b. white and milk chocolate
    c. nutty and plain

40. When filling molded chocolates, care must be taken to keep the filling
    a. at a temperature not to exceed 70°F
    b. above the melting point of the chocolate
    c. from overlapping the rims of the mold

Matching

Match each term in column 1 with its definition from column 2.

_____ 41. baking chocolate

_____ 42. Dutch processed cocoa
powder

_____ 43. bloom

_____ 44. couverture

_____ 45. white chocolate

_____ 46. seeding

_____ 47. ganache

_____ 48. nibs

_____ 49. chocolate liquor

_____ 50. tempering

a. three-stage process

b. unwanted affect to tempered
choclate

c. cooled chocolate liquor

d. one way to temper chocolate

e. uses an alkali

f. contains no chocolate liquor

g. truffle centers

h. inside the cocoa bean shell

i. a dark liquid

j. high-quality chocolate

CHAPTER 20

Dessert Sauces and Plating

After reading this chapter and successfully answering the following questions, you should be able to:

- Explain the four reasons to use dessert sauces.

- List the six basic types of dessert sauces.

- Use sauces to decorate a plate.

- Prepare the recipes in this chapter.

A Brief Outline of the Chapter

Key Terms

Explain these key terms.

1. caramel

2. coulis

3. crème anglaise

4. custard sauce

5. dessert sauce

6. sabayon or zabaglione

Fill in the Blanks

Use the most accurate word or phrase to complete each sentence.

7. The four main reasons for using dessert sauces are that they add moist-
ness, add _____, add complementary flavors, and create a
more attractive presentation.

8. The six main categories of dessert sauces include _____, _____, uncooked fruit sauces, cooked fruit sauces, chocolate sauces, and reduction sauces.

9. Strawberry purée that has been sweetened with sugar and then strained and drizzled over ice cream is a good example of an _____.

10. A common starch to add to thicken a cooked fruit sauce is _____.

11. If using a squirt bottle to apply the sauce when painting the plate, you should make certain that the sauce is _____.

12. It is important to use only _____ garnishes.

True or False

Identify each statement as true (T) or false (F).

_____ 13. A caramel sauce can be made by either the wet method or the dry method.

_____ 14. Reduction sauces, or glazes, are often applied using a pastry bag.

_____ 15. A zabaglione often contains Marsala wine.

_____ 16. When making crème anglaise, pasteurized eggs are usually used as the sauce is not cooked.

_____ 17. A coulis is an uncooked fruit sauce.

_____ 18. A cooked fruit sauce that is thickened with any starch can be frozen and thawed with no difficulty.

_____ 19. Vanilla is the usual flavoring for crème anglaise.

_____ 20. Ganache is the base of many chocolate sauces.

_____ 21. In a ganache, the ratio of cream to chocolate must always be 4 to 1.

_____ 22. In a reduction sauce, the object of cooking the base liquid and decreasing the volume is to increase the intensity of the flavor.

Multiple Choice

Identify the choice (a, b, or c) that most accurately answers the question or completes the statement.

23. In a cooked fruit sauce, the sauce is cooked briefly after the starch is added to thicken it but also to
    a. cook out the taste of the starch
    b. increase the intensity of the flavor
    c. decrease the intensity of the flavor

24. A basic sweet custard sauce that is a foundation for many desserts is
    a. coulis
    b. crème anglaise
    c. éclair

25. The higher the amount of cream to chocolate in a ganache, the
    a. thicker the sauce will be
    b. thinner the sauce will be
    c. denser the sauce will be

26. The two methods for making a caramel sauce are
    a. wet and dry
    b. sweet and sour
    c. hot and cool

27. The first procedure when preparing a chocolate sauce is to
    a. bring heavy cream, sugar, or butter to a boil
    b. whisk egg yolks
    c. strain the chocolate

28. The first procedure in making a reduction sauce is to
    a. cream butter and sugar
    b. whisk egg yolks
    c. combine the liquid and sugar in a heavy saucepan

29. The first procedure for making sabayon is to
    a. melt chocolate
    b. blend starch with water
    c. whisk egg yolks and sugar or sugar syrup over a hot water bath

30. A good tip when painting a plate with chocolate sauce is to
    a. chill the plate first
    b. have the plate at room temperature
    c. heat the plate

Matching

Match each term in column 1 with its definition from column 2.

_____ 31. sabayon

_____ 32. crème anglaise

_____ 33. chocolate sauce

_____ 34. reduction sauce base

_____ 35. ganache

_____ 36. starch

_____ 37. caramel sauce

_____ 38. garnish

_____ 39. coulis

_____ 40. squirt bottle

a. way to apply thin sauces

b. may be made from white or milk chocolate

c. added to cooked fruit sauces

d. cream and chocolate

e. zabaglione

f. a versatile custard base sauce

g. can be made by wet and dry methods

h. can add height to presentation

i. fruit juice, vinegar, apple cider, or alcohol

j. uncooked fruit sauce

CHAPTER 21

Healthy Baking

After reading this chapter and successfully answering the following questions, you should be able to:

- Modify some commonly used ingredients in recipes to healthier alternatives.

- Recognize the difference between more healthy and less healthy fats.

- List the three criteria for fat substitution.

- Identify the differences between sugar substitutes.

- Successfully substitute healthier ingredients for less healthy ingredients in recipes.

A Brief Outline of the Chapter

_____

Key Terms

Explain these key terms.

1. artificial sweeteners

2. saturated fats

3. unsaturated fats

4. trans-fats

_____

Fill in the Blanks

Use the most accurate word or phrase to complete each sentence.

5. White flour is _____ in that it has had the outer part of the grain or hull removed.

6. In adding alternatives to white flour, the object is to increase the _____ content of the final product.

7. If whole wheat pastry flour is substituted for white pastry flour, the _____ of the finished baked good will be darker.

8. In replacing all or part of the white all purpose flour in a recipe, consideration must be given to the reduction in _____ that may result as this substance gives baked goods their structure.

9. _____ in a recipe cannot be entirely eliminated as they contribute tenderness, flakiness, and flavor to the finished product.

10. Lard and butter are examples of _____ fats.

11. The type of fat in many processed foods that is partially hydrogenated, _____, have been determined to be very unhealthy and even linked to heart disease and cancer.

12. Sometimes a greater amount of _____ can be added to replace fat in baked goods because it also acts as a tenderizer.

13. In considering artificial sweeteners because they do not have the chemical makeup of _____, the chemical name for granulated sugar, they will react differently in baked goods.

14. The artificial sweetener closest to the chemical structure of sugar is _____. Although it can be substituted for sugar it can also cause unwanted changes in the texture of baked goods.

15. Examples of _____ like xylitol, mannitol, and sorbitol are artificial sweeteners often used in candies and gums.

True or False

Identify each statement as true (T) or false (F).

_____ 16. Saturated fats tend to be healthier than unsaturated fats.

_____ 17. Hydrogenated fats are liquid fats altered by having hydrogen atoms added.

_____ 18. Unsaturated fats tend to be liquid at room temperature.

_____ 19. Cocoa butter is a saturated fat.

_____ 20. Soy flour can be substituted for all purpose flour in that it provides the same gluten properties.

_____ 21. Trans-fats are extremely unhealthy.

_____ 22. All margarines are healthy alternatives to butter.

_____ 23. Reading the nutrition label on a food product will not tell you fat content.

_____ 24. Apple butter makes a good substitute for fat in some recipes.

_____ 25. Canola oil is a good example of a trans-fat.

_____ 26. The three criteria for fat substitution are the mixing method used, the form the fat takes, and the role of the fat in the recipe.

_____ 27. Lowering the fat content of a recipe will automatically decrease the calories.

_____ 28. Honey is a good example of an artificial sweetener.

_____ 29. All artificial sweeteners have the same basic chemical structure.

_____ 30. Aspartame is a carbohydrate as is granulated sugar.

Multiple Choice

Identify the choice (a, b, or c) that most accurately answers the question or completes the statement.

31. Although other modifications need to be made to the recipe to reduce fat, semisweet chocolate can often be replaced by
    a. chocolate chips
    b. unsweetened cocoa
    c. white chocolate

32. The following are examples of alcohol sugars:
    a. mannitol and sorbitol
    b. sucrose and glucose
    c. aspartame and sucralose

33. The disadvantage for bakers of this artificial sweetener is that it loses its sweetness when heated:
    a. sugar alcohols
    b. sucrose
    c. aspartame

34. When substituting egg whites for whole eggs in a recipe, there may be an unwanted result in
    a. the texture of the baked goods
    b. the calories of the baked goods
    c. higher cholesterol

35. The best substitute for reducing fat in a pie or tart crust is
    a. replace half the butter with a fat high in trans-fat
    a. replace half the butter with canola oil
    a. replace half the butter with trans-fat–free margarine

36. To lower the fat in a recipe that uses the creaming method, you can
    a. replace half the fat with canola oil
    b. replace half the fat with applesauce
    c. replace half the fat with fat-free milk

37. The healthy element found in chocolate is
    a. cocoa butter
    b. cocoa solids nonfat
    c. saturated fats

38. Chemical substances found in plants that can help repair human body cells are
    a. antioxidants
    b. saturated fats
    c. trans-fats

39. When replacing sour cream, yogurt, or milk in a recipe, it is safest to use
    a. the non-fat version
    b. the low-fat version
    c. the trans-fat version

40. When considering modifications to a recipe to lower either fat or sugar content, the best approach is
    a. a conservative one
    b. an aggressive one
    c. a fearless one

Matching

Match each term in column 1 with its definition from column 2.

_____ 41. xylitol

_____ 42. saccharin

_____ 43. trans-fat

_____ 44. fractionated oils

_____ 45. saturated fat

_____ 46. unsaturated fat

_____ 47. ground nuts

_____ 48. egg whites

_____ 49. sucrose

_____ 50. aspartame

a. made of two amino acids

b. granulated sugar

c. leaves a bitter aftertaste

d. olive oil

e. sugar alcohol

f. lard

g. replacement for whole eggs

h. replacement for all purpose flour

i. partially hydrogenated fats

j. made through a heating and cooling process

CHAPTER 22

Troubleshooting

After reading this chapter and successfully answering the following questions, you should be able to:

- Correct common mishaps or prevent them from happening the next time.

- Use common sense when baking, making sure to understand the methods, tools, and equipment needed.

A Brief Outline of the Chapter

VIII.   When something goes wrong with cookies

IX.   When something goes wrong while working with sugar

X.   When something goes wrong with frozen desserts

XI.   When something goes wrong with chocolate

Fill in the Blanks

Use the most accurate word or phrase to complete each sentence.

1. A good way to check the temperature of your oven is to place an
   _____ on the middle rack of your oven and
   compare the true temperature with what the oven indicates.

2. Cakes should be removed from the pans when they are barely
   _____ but not cold.

3. If the eggs are not properly tempered with the hot liquid in a custard,
   they may _____ and become lumpy.

4. Placing _____ directly on the top of a custard
   will prevent a thin _____ from forming.

5. By mistake, _____ liquid was added to the yeast dough, which
   created a bread that was coarse and had large holes.

6. A low-protein flour was incorrectly used to make the bread dough,
   which resulted in bread that _____.

7. A flour with an extremely high amount of protein was used, which
   resulted in bread with a crust that was _____.

8. The blueberry muffins turned out to be very tough, and it was discov-
   ered that the dough had been _____, causing
   too much gluten to form.

9. In a baking competition, one team used paper liners for their muffins,
   but the other team did not. The prize for the highest muffins went to
   the team that _____ paper liners.

10. The buttercream frosting for the cake looked grainy, and it was because
    the mixture had been _____.

True or False

Identify each statement as true (T) or false (F).

_____ 11. To assure that the bread dough will rise properly, a flour with a low-protein content is necessary.

_____ 12. If the pastry cream that was left in the refrigerator overnight has thinned too much to be used, it is probably because it was originally overcooked.

_____ 13. When too much baking soda is used in a cake batter, the finished cake may be too dark.

_____ 14. Too much fat in a cake batter can make a cake heavy and dense.

_____ 15. There is no way to repair a fudge frosting that has become too stiff to spread.

_____ 16. To help the sugar from crystallizing while making a caramel sauce, cream of tartar can be added at the beginning.

_____ 17. Too much alcohol added to a frozen dessert results in the item freezing solid.

_____ 18. It is best to place a lid on the double boiler when melting chocolate.

_____ 19. Ganache that has become grainy from overbeating cannot be saved.

_____ 20. Changing the fat in a cookie recipe can affect the amount that the cookie spreads during baking.

Multiple Choice

Identify the choice (a, b, or c) that most accurately answers the question or completes the statement.

21. The most common cause of a bloom appearing on chocolate is from
    a. improper storage
    b. too high a cocoa butter content
    c. too low a coca butter content

22. Washing down the sides of the pan while making a caramel sauce prevents
    a. the temperature from getting too high
    b. the mixture from getting too dry
    c. crystallization from occurring

23. The following are all causes of cookies spreading too much and flattening except
    a. too much sugar
    b. too little flour
    c. oven temperature was too high

24. The result in the buttercream frosting when the butter is added to the hot sugar and egg mixture is that
    a. it becomes grainy
    b. it does not thicken
    c. it is too stiff to spread

25. The result in the buttercream frosting when the butter is too cold when it is added to the sugar and egg mixture is that
    a. it is grainy
    b. it looks curdled
    c. it is too thick

26. A tough cake may be due to
    a. improper mixing
    b. using a flour with very low protein
    c. using too much sugar

27. When too little fat is used for a pie crust, the result will be
    a. a crust that falls apart
    b. a crust that is tough
    c. a bottom crust that is soggy

28. One simple way to remove lumps in a custard is to
    a. strain them out
    b. add an acid
    c. add cold milk while whisking

29. In blind baking a pie shell, docking or stippling is done to prevent
    a. sogginess
    b. overbaking
    c. shrinking

30. For a cake-like texture in a muffin, the mixing method of choice is the
    a. creaming method
    b. egg-foam method
    c. one-stage method

Matching

Match the result in column 1 with the reason from column 2.

_____ 31. Custard was lumpy.

_____ 32. Bread would not rise.

_____ 33. Bread broke apart.

_____ 34. Pie crust was tough.

_____ 35. Muffins were not high enough.

_____ 36. Angel food cake collapsed.

_____ 37. Cookies flattened.

_____ 38. Sugar syrup boiled over.

_____ 39. Sherbet lacked flavor.

_____ 40. Bloom appeared on the chocolate.

a. Oven was set too high.

b. Pan was too small.

c. Yeast was dead.

d. Insufficient flavoring was added.

e. Paper liners were used.

f. Excess air was incorporated.

g. Storage was improper.

h. Eggs were not properly tempered.

i. Flour had too much protein.

j. Flour had too little protein.

Appendix A:
Weights and Measurements

Weights and Measurements

This section has been written to give you instant access to a table of weights and measurements when you are entering recipes. The weights and measurements are only approximate: you can never be precise for many reasons:

1. The moisture contents of products vary constantly.
2. Sizes of individual pieces or particles will vary from container to container.
3. The exact weight of a gallon or a pound of product is seldom a convenient round number.
4. It would be impractical to say that a pint of water is 1 9/10 cups. It is simpler just to say 2 cups.
5. Products containing moisture become lighter as they dry out.
6. Wet products containing sugar become heavier when the moisture evaporates and they become thicker.
7. A cup of flour could weigh 4 ounces. If you sift it, it may weigh less.
8. Any measurement such as a "Level" teaspoon or "Level" cupful is seldom exactly accurate.

Gram Weight Conversion Table

OUNCES	GRAMS	POUNDS	GRAMS
1	28.35	1	453.6
2	56.70	2	907.2
3	85.05	2.5	1134.0
4	113.40	3	1136.8
5	141.75	4	1814.4
6	170.10	5	2268.0
7	198.45	6	2721.6
8	226.80	7	3175.2
9	255.15	8	3628.8
10	283.50	9	4082.4
11	311.85	10	4536.0
12	340.20	15	6804.0
13	368.55	20	9072.0
14	396.90	25	11340.0
15	425.25	30	13608.0
16	453.60	35	15876.0

(*Note:* When you know the weight in ounces, multiply by 28.35 to find the grams. Divide the grams by 453.6 to find the pounds.)

Metric Size Fluid Equivalents

METRIC	U.S. FL OZ	3/4 OZ	1 OZ	1-1/8 OZ	1-1/4 OZ	1-1/2 OZ	CLOSEST PREVIOUS CONTAINER U.S. OZ
1.75 Liter	59.2	78.9	59.2	52.6	47.4	39.5	½ Gal. = 64 OZ
1.0 Liter	33.8	45.1	33.8	30.0	27.0	22.5	Qt. = 32 OZ
750 Milliliters	25.4	33.9	25.4	22.6	20.3	16.9	5th = 25.6 OZ
500 Milliliters	16.9	22.5	16.9	15.0	13.5	11.3	Pt = 16 OZ
200 Milliliters	6.8	9.1	6.8	6.0	5.4	4.5	½ Pt. = 8 OZ
50 Milliliters	1.7						Miniature = 1.6 OZ

Equivalent Measures for Fluids

(*Note:* This table gives measurement/weight equivalencies for water. Other liquids may vary.)

3 teaspoons	= 1 tablespoon
16 tablespoons	= 1 cup
28.35 grams	= 1 ounce
1 cup	= ½ pint
2 cups	= 1 pint
2 pints	= 1 quart
4 quarts	= 1 gallon
768 teaspoons	= 1 gallon
1 lb. (water)	= 16 fluid ounces
1 lb. (water)	= 1 fluid pint
2 lb. (water)	= 1 fluid quart

Volume Conversions for Recipe Writing

	TEASPOON	TABLESPOON	QUART	CUP
1 teaspoon	1.0	0.333333	0.0052083	0.020833
1 tablespoon	3.0	1.0	0.015625	0.062500
1 cup	48.0	16.0	.25	1.0
1 pint	96.0	32.0	.50	2.0
1 quart	192.0	64.0	1.0	4.0
1 gallon	768.0	256.0	4.0	16.0

Pounds and Ounces to Grams

OUNCES	GRAMS	POUNDS	GRAMS
1	28.35	1	453.60
5	141.75	5	2268
10	283.50	10	4536
12	340.20	25	11340
16	453.60	50	22680

Teaspoons and Tablespoons

TEASPOONS	TABLESPOONS
3 = 0.5 ounce	1 = 3 teaspoons
6 = 1 ounce	2 = 1 ounce
48 = 1 cup	4 = 0.25 cup
96 = 1 pint	8 = 0.5 cup
192 = 1 quart	16 = 1 cup
960 = 5 quarts	128 = ½ gallon
768 = 1 gallon	256 = 1 gallon

Decimal Equivalents of Common Fractions

FRACTION	ROUNDED TO 3 PLACES	ROUNDED TO 2 PLACES
5/6	0.833	0.83
4/5	0.8	0.8
3/4	0.75	0.75
2/3	0.667	0.67
5/8	0.625	0.63
3/5	0.6	0.6
1/2	0.5	0.5
1/3	0.333	0.33
1/4	0.25	0.25
1/5	0.2	0.2
1/6	0.167	0.17
1/8	0.125	0.13
1/10	0.1	0.1
1/12	0.083	0.08
1/16	0.063	0.06
1/25	0.04	0.04

Fahrenheit to Celsius Conversion (Approximate)

FAHRENHEIT	CELSIUS
32°F	0°C
50°F	10°C
68°F	20°C
86°F	30°C
100°F	40°C
115°F	45°C
120°F	50°C
130°F	55°C
140°F	60°C
160°F	70°C
170°F	75°C
180°F	80°C
185°F	85°C
195°F	90°C
200°F	95°C
212°F	100°C
230°F	110°C
250°F	120°C
265°F	130°C
285°F	140°C
300°F	150°C
325°F	165°C
350°F	175°C
360°F	180°C
375°F	190°C
400°F	200°C
425°F	220°C
450°F	230°C
485°F	250°C
500°F	260°C
575°F	300°C

Sizes and Capacities of Scoops

NUMBER ON SCOOP	LEVEL MEASURE
6	2/3 cup
8	1/2 cup
10	3/8 cup
12	1/3 cup
16	1/4 cup
20	3-1/3 tablespoons
24	2-2/3 tablespoons
30	2 tablespoons
40	1-2/3 tablespoons
50	3-3/4 teaspoons
60	3-1/4 teaspoons
70	2-3/4 teaspoons
100	2 teaspoons

Sizes and Capacities of Measuring/Serving Spoons

SIZE OF MEASURING/SERVING SPOON	APPROXIMATE MEASURE
2 ounces	1/4 cup
3 ounces	3/8 cup
4 ounces	1/2 cup
6 ounces	3/4 cup
8 ounces	1 cup

Appendix B:
Canned Goods

Common Can Sizes and Their Appropriate Contents

CAN SIZE	PRINCIPAL PRODUCTS
No. 5 squat	
75 oz. squat	
No. 10	Institution size—fruits, vegetables, and some other foods
No. 3 cyl	Institution size—condensed soups, some vegetables. Meat and poultry products. Economy family size—fruit and vegetable juices
No. 2½	Family size—fruits, some vegetables
No. 2	Family size—juices, ready-to-serve soups, and some fruits
No. 303	Small cans—fruits, vegetables, some meat and poultry products, and ready-to-serve soups
No. 300	Small cans—some fruits and meat products
No. 2 (vacuum)	Principally for vacuum pack corn
No. 1 (picnic)	Small cans—condensed soups, some fruits, vegetables, meat, and fish
8 oz.	Small cans—ready-to-serve soups, fruits, and vegetables
6 oz.	

Common Can Sizes and Their Approximate Weights and Volumes

Volumes represent total water capacity of the can. Actual volume of the pack would depend upon the contents and the head space from the fluid level to top of can.

AVERAGE CAN SIZE	AVERAGE VOLUME	CANS PER FLUID	APPROX CUPS	CASE	WEIGHT
75 squat				6	4 lb 11 oz
No. 5	56			6	4 lb 2 oz
No. 10	105.1	99 to 117	12 to 13	6	6 lb 9 oz
No. 3 cyl	49.6	51 or 46	5¾	12	46 fl oz
No. 2½	28.55	27 to 29	3½	24	12 oz
No. 2	19.7	20 or 18	2½	24	1 lb 13 oz
No. 303	16.2	16 or 17	2	24 or 36	1 lb
No. 300	14.6	14 or 16	1¾	24	15½ oz
No. 2 (vacuum)		12 fl oz	1½	24	
No. 1 (picnic)		10½ fl oz	1¼	48	
8 oz.	8.3	8 fl oz	1	48 or 72	8 oz
6 oz.	5.8	6 fl oz	¾	48	6 oz

GLOSSARY

acetate A clear, thin, flexible plastic sold in rolls, sheets, and strips used in the molding of chocolate, protecting the sides of a cake, or coating the sides of a cake with chocolate.

Acetobacillus A species of bacteria that exist in a sour dough starter. These bacteria give off acetic acid, providing a slightly tangy taste to the finished bread.

acid A substance that tastes sour like lemon juice and has a pH of less than 7.0.

active dry yeast Fresh yeast that has been dehydrated. It is more concentrated than fresh yeast and has a longer shelf life. See *yeast*.

agar A gelatin-like stabilizer and thickener that is derived from a type of seaweed known as red algae. Also referred to as *agar agar*.

agar agar See *agar*.

air cell The pocket of air that forms at the larger end of an egg.

all-purpose flour Flour made from a combination of hard and soft wheats so as to be suited for all purposes. Protein levels may vary depending on where it is milled.

almond paste Almonds and sugar ground into a fine paste.

alpha-amylase An enzyme in raw egg yolks that feeds on starch, causing it to break down.

amino acids The building blocks of protein.

ammonium carbonate   Also known as ammonium bicarbonate; a chemical leavener that, in the presence of moisture and heat, reacts to produce ammonia, carbon dioxide gas, and water.

angel food cake   An egg foam cake that uses only egg whites and is virtually fat free.

artificial sweeteners   Substitutes for sugar (sucrose) that are artificial or man-made in the laboratory. They do not tend to raise blood sugar levels. Artificial sweeteners may not have all of the properties of sugar and may not be suitable for baking.

artisan breads   Breads that are prepared by bakers who manipulate the dough with their hands with great care and skill using traditional methods.

ascorbic acid   Also known as vitamin C. It is added to flour by the miller to improve gluten quality.

autolyse   A short rest given to a yeast dough before kneading has begun. This rest helps gluten develop properly.

baguette pan   Also known as a *French bread pan*; a long metal pan formed into half cylinders that are joined together side by side. Frequently small holes are placed in the metal to allow for better air circulation and a crisper crust.

bain marie (ban mah-ree)   A French term for a hot water bath that can be used either to warm or melt ingredients or to surround custards in the oven to ensure even cooking. When used as a double boiler, with a bowl placed over a pot of simmering water, it can melt chocolate or warm other delicate ingredients. It can also be used to ensure a more gentle, even cooking for custard desserts like crème brûlée, crème caramel, or cheesecakes. Ramekins full of crème brûlée custard or a springform pan filled with cheesecake batter can be placed in a larger rectangular pan, which is then filled halfway with hot water. The water surrounds the custard, providing it with a consistent temperature, which prevents the eggs in the custard from curdling.

baked custard   A mixture of eggs and milk or cream, sometimes with additional ingredients, which is poured into a container and baked in the oven until thickened.

baker's peel   A flat shovel-like blade with a long handle used to transfer bread or pizza dough into or out of an oven when it will not be baked directly on a sheet pan. Peels can be made of wood, stainless steel, or a combination of the two.

baker's percentages   A system used by professional bakers (especially bread bakers) in large commercial operations that involves percentages to express formulas in a simple way.

baking   The act of placing a food such as a dough or other unbaked pastry in the oven where dry heat cooks the food. The term *baking* is generally applied to cakes, pies, yeast breads, cookies, and quick breads. Baking helps proteins and starches to set and doughs to rise.

baking chocolate   Chocolate liquor in a solid state; also referred to as *unsweetened chocolate* or bitter chocolate.

baking powder   A chemical leavener that contains sodium bicarbonate (baking soda) and at least one acid that is used to help baked goods such as quick breads and cakes to rise. See *double-acting baking powder* or *single-acting baking powder.*

baking soda   A chemical leavener known as sodium bicarbonate that, when combined with an acid in the presence of moisture, forms carbon dioxide gas and is used to help baked goods to rise.

banneton   A woven basket made of coiled reed or willow of various shapes and sizes, sometimes lined with cloth. Rustic and hearth-type bread doughs are allowed to rise in them, imparting an attractive pattern onto the dough before it is baked.

bar cookie   The category of cookie preparation wherein a stiff dough is shaped into long bars or logs, baked, and sliced.

barm   An English whole wheat sourdough starter made from wild yeast.

base   A substance with a pH of greater than 7.0 that neutralizes acid to produce a salt, for example, baking soda. Also known as an alkali.

bench scraper   A small tool consisting of a rectangular blade attached to a wooden or plastic handle. It is used to cut and scale pieces of dough and to clean work surfaces by scraping it against a table to loosen pieces of dough or flour. Also known as a *dough scraper*.

benching   A stage in yeast dough production in which the dough is scaled into pieces, covered, and allowed to rest for a short period of time before being shaped. See *resting*.

biga (BEE-gah)   An Italian word for a thick sponge starter.

biscuit method   A mixing method for quick breads that resembles the method to make pâte brisée or flaky pie crust by first cutting cold fat into dry ingredients and then adding liquid ingredients.

blind baking   When a pie or tart shell is baked with nothing in it. The shell is lined with parchment paper and pie weights or dried beans to keep its shape during baking. Used most often for pies whose filling is prepared separately and requires no further cooking.

bleaching   The process whereby newly milled flour is exposed to a bleach such as chlorine gas or benzoyl peroxide to whiten it.

bloom   (1) When tempered chocolate that has been exposed to temperature variations and/or humidity, it develops a whitish-gray spotty outer coating. (2) When a cold liquid is added to powdered gelatin, the absorbed liquid causes it to soften and swell so that it appears to be blooming. (3) Also refers to the Bloom rating, a system used to show how firm or strong a specific type of gelatin is.

bowl scraper   A small flexible piece of plastic used to scrape around the inside of a mixing bowl to loosen doughs or stiff batters for easier removal.

bran   The hard outside covering of the wheat kernel. Also known as the *hull*.

bread flour   The general term for flour that is milled from wheats having a higher protein content and generally used for bread making.

breaking   In the milling of flour, when special machines crack or break open wheat kernels to separate them into their component parts.

buttercream   A frosting consisting of butter or vegetable shortening, granulated sugar, corn syrup, or confectioners' sugar, and whole eggs, yolks, or egg whites. There are different types of buttercreams, including both uncooked (e.g., simple buttercream) and cooked variations (e.g., French, Italian, and pastry cream-based buttercreams).

buttermilk   Buttermilk traditionally referred to the liquid left over after cream was made into butter. Presently, the buttermilk on the market is cultured and refers to skim or low-fat cow's milk that is treated with harmless bacteria, giving it a thick consistency and a sour taste. It has a milk fat content between 0.5 and 3 percent. Because of its acidic nature, it reacts well with baking soda, neutralizing it to form carbon dioxide gas to leaven cakes and quick breads. Because of its acidity and low fat content, buttermilk has a longer shelf life than regular cow's milk. Also known as cultured buttermilk.

cacao beans (kah-KAH-oh)   The fruit of the *Theobroma cacao* tree from which chocolate is derived. Also referred to as cocoa beans.

cake   A sweet, tender, moist baked pastry that is sometimes filled and frosted.

cake flour   A flour milled from soft wheats that is typically bleached and used for only the most tender cakes and pastries.

caramelized sugar   Sugar that is cooked to within the temperature interval of 320° to 350°F (160° to 177°C), which causes the sugar to develop a brown color and a rich intense flavor.

cardamom   A fragrant pod related to the ginger family used in Middle Eastern and Indian dishes. It gives a pleasant, pungent aroma to Danish dough.

carrageenan   A type of seaweed from Ireland that is similar to agar that is used to thicken foods containing dairy products.

carryover cooking   Refers to the process in which cooking continues for a short period of time, even though the heat source has been removed.

casein (kay seen)   A protein in milk and other dairy products, which when exposed to air, forms a crusty skin.

chalaza (kuh-LAY-zah)   The white stringy material that anchors the yolk in the center of the egg.

chef   A chef is the beginning stages of a sourdough starter. It is also known as a *seed culture*. After a period of time, the starter becomes healthy enough to bake bread.

chemical leaveners   Refer to *baking powder*, *baking soda*, and *ammonium carbonate*, which are chemicals that react with liquid ingredients upon mixing and the heat of the oven to produce carbon dioxide gas that leavens baked goods.

chiffon cake   A type of egg foam cake containing a liquid fat and a chemical leavener. These cakes tend to be moister than typical sponge cakes and are baked in a tube pan.

chocolate confectionery frostings   A category of chocolate frosting based on two types of chocolate confections: fudge and truffles.

chocolate liquor   The dark liquidy paste created when chocolate nibs are crushed. Once cooled into bricks or disks, it is known as unsweetened chocolate, baking chocolate, or bitter chocolate.

choux paste   See *éclair paste*.

churn-frozen desserts   Frozen desserts that are churned or stirred as they freeze. Churn-frozen desserts include ice cream, sorbets, and sherbets.

cinnamon   A spice originating from the inner bark of an evergreen laurel tree, cinnamon has a sweet, spicy aroma. It is ground or sold as curls of bark called sticks or quills. It is one of the most popular spices in a baker's kitchen. It is used extensively in pies, cakes, cookies, and fruit desserts.

classifying   The final stage of flour milling wherein the particles of flour are categorized by size.

clear flour   The particles of flour from the outermost layers of the endosperm. It is the flour that remains after patent flour is removed.

clove   A spice originating from the dried, unopened flower buds of the tropical evergreen tree called the clove tree. The buds are sold whole or ground into a deep mahogany powder. Clove is pungent, yet sweet, which is ideal in various cakes, cookies, pies, and other desserts.

coagulation   When proteins are heated and moisture gets trapped between each protein coil, the protein forms a network that produces a thick, gel-like structure. An example of coagulation is when egg proteins thicken a custard.

cocoa butter   The saturated fat that is naturally present in chocolate liquor that gives chocolate its characteristic velvety texture.

cocoa powder   The dry powdery residue that remains when cocoa butter is removed from chocolate liquor.

coconut   Coconut is the fruit grown on tropical palm trees. It consists of a hard, fibrous, outer brown shell, which when cracked open, yields a white, hard flesh with a center filled with a milky liquid called coconut water or coconut milk. Coconut is sold whole in the shell, shredded and sweetened or unsweetened, flaked, grated, and ground.

cold-water test   A test to determine whether a sugar syrup is done by dropping some syrup into a glass of cold water to see how easily it can be gathered into a ball.

compound coating   Chocolate that contains little or no cocoa butter and is used to coat candies. It has a longer shelf life but tends to be of lesser quality.

conching (KONCH-eng)   Part of the chocolate making process when chocolate liquor is rotated and stirred with blades to develop flavor and texture.

conditioning   A general term used for when a miller adds certain chemicals such as ascorbic acid or diastase to newly milled flour to help it produce better gluten, provide the best food for yeast, and overall improve the qualities of the finished baked good.

cookie   A diverse group of small, sweet cakes or pastries that are described and categorized by how the dough is prepared for baking.

couche   A piece of heavy canvas or linen in which a yeast bread can be nestled in order to hold its shape during the proofing process.

coulis   An uncooked fruit sauce of fresh or frozen puréed fruit that is sweetened and strained.

coupler   A coupler is a plastic cone-shaped tube that is used to allow various pastry tips to fit onto a pastry bag to facilitate the piping of frostings and batters. The coupler allows tips to be changed during decorating without having to change pastry bags.

couverture   High-quality chocolate made with at least 32% cocoa butter that is used in baked goods or to make candy bars, or to coat candies and create decorations for all sorts of pastries. Couverture comes in milk, semisweet, or white varieties.

cream cheese   Cream cheese is a soft, spreadable, unaged cheese that is cultured with bacteria to give it a slight tang. It is used in cheesecakes, cookies, pastry doughs, and pie crusts.

cream of coconut   Cream of coconut is also referred to as coconut cream. Cream of coconut is a thick liquid that is intensely rich in coconut flavor and is made from the liquid that rises to the top of the coconut milk, sugar, and other thickening agents. It is used in many desserts and alcoholic beverages.

cream of tartar   Chemically known as potassium hydrogen tartrate, an acidic salt of tartaric acid, cream of tartar is formed during the wine making process and is deposited on the inside of wine barrels. It is used to make meringues more stable and to help prevent candies from crystallizing.

creaming   See *creaming method*.

creaming method   A method of mixing in which granulated or brown sugar is mixed with a softened, solid fat using the paddle attachment of an electric mixer until it is light and fluffy. Air is incorporated into the fat and is instrumental in aiding the leavening process.

crème anglaise (krehm ahn-GLEHZ)   A sweet, French stirred custard sauce used as a dessert sauce or as a base for frozen desserts. It is made from egg yolks, sugar, milk, half-and-half, or heavy cream; and various flavorings. A typical crème anglaise is flavored with vanilla and is referred to as a vanilla custard sauce.

croissant cutter   A tool used to cut croissants from croissant dough that resembles a short rolling pin with a cut out triangular shape between the handles.

croissant dough   A laminated yeast dough that is formed when butter is encased in a base dough containing yeast, then rolled and folded repeatedly to make multiple thin layers. Traditionally the dough is cut into triangles, shaped into crescents and baked. The finished rolls are known as *croissants*.

crystallization   When particles of a pure substance such as sugar (sucrose) form a repeated shape and are packed closely together to form crystals.

curdling   When egg proteins clump together because they are heated for too long and at too high a temperature.

cutting   The technique used to combine fat and dry ingredients until the pieces of fat have been reduced to a desired size. This is accomplished using a pastry blender, food processor, or an electric mixer. This mixing technique is used in the flaky pie crust method of preparing crusts for pies and tarts and in the biscuit method of mixing to prepare quick breads.

Danish dough   The dough from which Danish pastry is made. A rich yeast dough in which fat is enclosed and then rolled and folded repeatedly to make multiple thin layers. It is baked into flaky breakfast pastries and coffee cakes.

degassing   See *punching*.

denature   When proteins such as eggs are heated, beaten, or acidified, causing the protein strands or coils to straighten out or break apart.

desem   A natural starter using only whole wheat flour that produces very dense bread with little acidity.

detrempe (day-trup-eh)   The French term for a base dough used in laminated doughs.

diastase   An enzyme that acts as a catalyst to help starches within flour break down into sugars.

docking   Piercing the bottom and sides of a raw pie or tart shell with a fork or special instrument to prevent the pastry from puffing up and shrinking in the oven. Also referred to as *stippling*.

double-acting baking powder   A chemical leavener that requires both moisture and heat to produce carbon dioxide gas that is used to leaven baked goods.

dough hook   A tool on an electric mixer shaped like a hook that is used to mix yeast doughs. The dough hook helps to simulate the kneading process.

dough scraper   See *bench scraper*.

drop cookie   The category of cookie preparation wherein a dough is dropped from a spoon onto a sheet pan and then baked.

dry method   A method of preparing caramelized sugar by heating sugar in a heavy pan without any water. An acid may be added to prevent crystallization.

Dutch processed cocoa powder   Cocoa powder that has been treated with an alkali to neutralize its acidity.

éclair paste   Also known as *choux paste* or *pâte à choux*. A steam-leavened specialty dough used to prepare cream puffs and éclairs.

egg foam   When air is beaten into whole eggs or egg whites forming a foam that can leaven baked goods.

egg-foam cakes   The category of cakes that uses air beaten into eggs (an egg foam) to leaven them.

egg-foam frostings   Also known as *boiled frostings*, egg-foam frostings consist of Italian and Swiss meringues. Gelatin may also be added.

electric mixer   A piece of equipment used frequently in baking consisting of a bowl fitted to a motor on which three standard attachments are included: a paddle for mixing, a whip for beating, and a dough hook for kneading. Other attachments are available. Electric mixers can be small enough to fit on a work table or large enough to be permanently attached to the floor.

emulsified shortening   A solid fat containing emulsifiers that are able to hold a large amount of liquid to keep cake batters uniformly combined and emulsified. It is generally used for high ratio cakes.

emulsifying agent   A food (e.g., egg yolks) or a food additive that allows two immiscible liquids to stay uniformly mixed together without separating.

emulsion   A uniform mixture of two unmixable substances such as fat and water-based ingredients to create a homogeneous mixture that will not separate.

endosperm   The largest portion of a wheat kernel located under the bran layer. It is used to make white flour.

enrichment   The process of adding certain vitamins and minerals back into the flour that were lost during the milling process.

enzyme   A protein that speeds up a chemical reaction. An example of an enzyme is *alpha-amylase*, which breaks down starch into sugar.

evaporated milk   Unsweetened whole milk from which 60 percent of the water has been evaporated. It contains at least 7.5 percent milkfat and is sold in cans. It is used in confections, frostings, and baked goods.

evaporated skim milk   Unsweetened skim milk from which 60 percent of the water has been evaporated. It is the same as evaporated milk except skim milk is used. It contains less than 0.5 percent milkfat.

false-bottom tart pan   A tart pan used for making a fluted pastry crust that can be filled with sweet or savory fillings that has a removable bottom for easy removal. Tart pans have different shapes, such as round, square, or rectangular. The round tart pans have varying diameters.

false-bottom tart ring base   A thin metal circle used as the removable bottom of a round false-bottom tart pan and which can be used to help separate two cake layers.

fast-rising dry yeast   See *instant active dry yeast.*

fat bloom   When crystals of fat travel to the surface of chocolate and recrystallize on the outside to form a whitish coating.

fermentation   (1) The process of yeast eating sugar and converting it to carbon dioxide gas and alcohol. (2) In yeast dough production, the first rise of a yeast dough in which carbon dioxide gases are produced and become trapped in a network of gluten.

Flexipan   The brand name for a type of baking pan made from flexible silicone that can withstand a wide range of temperatures and has a permanent, nonstick surface, much like a silicone baking mat. Flexipans come in a wide variety of shapes and sizes and need to be baked on a rigid surface such as a half or full-size sheet pan.

flour   Grains or nuts ground or milled into various degrees of fineness to create a meal or powder.

folding   A very gentle method of blending lighter, air-filled ingredients into heavier batters, usually with a rubber spatula, without losing air volume.

fondant   See *poured fondant* or *rolled fondant.*

food processor   A bowl with a blade inside it attached to an electric motor that is used for chopping and blending.

formula   The term used by professional bakers (especially bread bakers) for a recipe.

four-fold or **bookfold turn**   When a laminated dough is folded like a book to produce many layers of fat and dough.

French bread pan   See *baguette pan.*

French meringue   A mixture of egg whites and sugar beaten to stiff peaks, also known as a *common meringue*, it is the simplest type of meringue.

fresh currants   The small, shiny berries that grow on a prickly shrub in clusters like miniature grapes. They are tart in flavor and are red, black, or golden (also known as white). They are used to make jellies and fillings for cakes and pastries, and are used for garnishes. Not to be confused with dried currants.

fresh or **compressed yeast**   Fresh yeast mixed with a starch that is portioned or compacted into a small cube. Fresh yeast has a short shelf life and must be kept refrigerated. See *yeast.*

friction   The heat energy transferred to a yeast dough through the act of mixing.

friction factor   The difference between the temperature of the dough before and after mixing. The friction factor can be calculated for a particular mixer by taking the sum of the temperature of the room, the flour, and the water and subtracting it from the actual dough temperature multiplied by three.

frosting   A sweet topping or covering used to fill or coat the top or sides of cakes, cookies, and other pastries.

fudge-style frostings   A frosting based on fudge candy beginning with a boiled sugar syrup to which butter and flavorings are added.

galette   A French word referring to a free-form tart usually filled with fruit.

ganache   (ga-nosh) A versatile mixture of cream, chocolate, and flavorings used to make sauces, glazes, frostings, and candies.

ganache-style frostings   A rich frosting consisting of a mixture of simmering cream to which chocolate has been added and allowed to cool and thicken.

gelatin   A stabilizer that helps foods form a gel-like consistency, giving structure to desserts. It is derived from animal connective tissue or bones, or from plants. Left to bloom in a cold liquid, gelatin is then dissolved in hot ingredients or over a hot water bath and then chilled. It is available in sheets or as a granular powder.

gelatinization   The process that starch granules go through to ultimately thicken a liquid. Gelatinization occurs when a starch and a liquid are heated, the starch absorbs the liquid, then the starch swells and ultimately thickens the liquid.

gelatinization of starches   During the baking process, starches within the flour absorb moisture from the dough, swell, and become firm.

gelation   When gelatin firms up or sets up to become a solid-like gel.

genoise (jehn-waahz)   A type of egg-foam cake known as a *whole egg foam* or *sponge cake* in which whole eggs are warmed and beaten with sugar until thick and then folded into dry ingredients, usually with the addition of melted butter.

germ   The smallest part of the wheat kernel or other grain. The germ regenerates the plant and is the only part containing fat.

ginger   A spice that originates from a flowering tropical plant native to
   China. Ginger is part of an underground root system called a *rhizome*
   that grows horizontally. It has short, finger-like projections covered in a
   light brown skin. Once peeled, ginger can be sold fresh, pickled,
   ground, or crystallized. It has a sweet, almost peppery taste and is used
   extensively in baked goods.

glaze   A category of frosting used as a thin coating for cake layers, tarts,
   cookies, yeast breads, and coffee cakes, consisting of sugar syrups or
   thinned and melted preserves or jams.

gliadin   See *gluten*.

glutathione   A protein fragment (amino acid) found in milk and active dry
   yeast that weakens gluten in yeast doughs.

gluten   The network of fibers that is created when two proteins in wheat—
   glutenin and gliadin—are mixed with water. This web of fibers keeps
   gases trapped, causing baked goods to rise and providing strength and
   structure.

glutenin   See *gluten*.

granita (grah-nee-TAH)   An Italian sweetened, flavored, slushy ice that is
   scraped after freezing and scooped into glasses. It is served as a light
   dessert or a palate cleanser for in between courses of a meal. See *granité*.

granité (grah-nee-TAY)   The French version of a *granita*.

grater   A rectangular strip or box of metal with sharp holes of varying sizes
   cut out. Foods are passed up and down to allow small slivers to fall
   through to the other side. Graters can be used to remove the outer peel
   from citrus fruits, or to shred cheese, vegetables, or chocolate. It is similar to a microplane zester.

gum arabic   A gelatin-like stabilizer derived from the sticky sap of a tree
   that grows in Africa. Gum arabic is used for stabilizing emulsions in
   frostings and fillings.

gum tragacanth A gum derived from a Middle Eastern shrub. It is used as a stabilizer for gum paste decorations and flowers when it is mixed into fondant. The decorations have the feel of bone china when they dry.

gums The collective term for gelatin-like thickeners and stabilizers that are derived from plants, also known as vegetable gums. Examples of gums include *agar, carrageenan, gum arabic,* and *gum tragacanth.*

hard wheat Wheat that is grown in harsher climates and contains greater amounts of protein and lower amounts of starch.

heavy cream Cream that is pasteurized but not homogenized with a milk-fat content of 36 to 40 percent and used to prepare whipped cream because of its ability to hold air. It is also used in frozen desserts such as ice cream and for ganache and in caramel and other rich sauces.

high-gluten flour Flour containing a high level of protein (approximately 14 percent) used in yeast breads where a chewier texture is desired. High-gluten flour can also be used in combination with other flours that may lack gluten-forming proteins to strengthen them.

high-ratio cake A cake that contains more sugar than flour by weight.

homogeneous When different ingredients are thoroughly mixed together to form a uniformly blended mixture.

homogenization A process whereby fat blobs are broken down into tiny particles so that they stay evenly dispersed in milk.

hull The hard outside covering of the wheat kernel or other grain. See *bran.*

hydrogenation When hydrogen is added to liquid fats such as oils to chemically and physically alter them to a solid form.

hygroscopic When a substance absorbs water from the air, keeping the substance moist. Sugar is an example of a hygroscopic substance.

icing See *frosting.*

icing comb   A rectangular or triangular piece of hard plastic or metal in which grooves or ridges have been cut out at regular intervals. It is dragged along the sides or top of a newly frosted cake where it leaves designs imprinted onto the frosting.

immiscible liquids   Two unmixable liquids (fat based and water based) that do not naturally stay blended together. Examples of two immiscible liquids are oil and water.

instant active dry yeast   A type of yeast that absorbs water instantly. It can be mixed directly in with the dry ingredients of a recipe. It produces more carbon dioxide gas per yeast cell, so a smaller amount can be used in yeast breads as compared to active dry yeast. Also known as *fast-rising dry yeast*. See *yeast*.

interfering agents   Ingredients added to a sugar syrup to keep sugar molecules from recrystallizing by preventing the sugar crystals from joining together. An example of an interfering agent is an acid or a different type of sugar such as corn syrup.

inversion   The process of preventing crystallization by adding an acid while heating a sugar syrup to break down the existing sugar into its component parts, thereby creating an impure state and controlling crystallization.

invert sugar   An invert sugar is sucrose that is chemically broken down into its two component parts (glucose and fructose) through the process of inversion.

Italian meringue   A type of meringue in which a hot sugar syrup is beaten into egg whites. Italian meringues are the most stable type of meringue.

kneading   A stage in yeast dough production in which the dough is pushed against a work surface and folded over onto itself until a smooth, elastic dough has developed. Gluten is developed during this process.

Lactobacillus   A species of bacteria that exist in a sour dough starter. These bacteria give off lactic acid, providing a slightly tangy taste to the finished bread.

laminated doughs   Rich doughs with or without yeast in which fat has been incorporated through a series of folds or turns. When baked, laminated doughs form hundreds of layers of flaky pastry such as in croissants, Danish pastry, and puff pastry. Also known as *rolled-in doughs*.

lean doughs   Yeast doughs that use little or no fat or sugar. They include breads that are prepared using few ingredients—French bread, Italian breads, pizza dough—and tend to have hard crusts.

leaveners   Ingredients that are added to batter and dough to help them rise. They include baking powder, baking soda, ammonium carbonate, yeast, and eggs. Natural leavening agents such as air and steam are added through the act of mixing and the addition of water-based ingredients.

levain   A sourdough starter made from wild yeast that is used to leaven a sourdough bread known as *pain au levain*.

litchis   Also spelled lychees. Litchis are small, round fruit native to Asia with a rough, leathery, inedible red skin and a delicate white flesh that encases a brown seed. The fruit has a light, sweet flavor with overtones of flowers, much like a perfume. It is available fresh or canned.

loaf pan   A rectangular pan, in different sizes, with high sides, generally used to bake quick breads and pound cakes.

low-fat milk   Milk from a cow that has had some of the milkfat removed so that it contains anywhere from 0.5 to 3 percent milkfat.

lychees   See *litchis*.

Maillard reaction   A reaction between amino acids and sugars that occurs between 300° and 500°F (149° and 260°C) causing the outer crust of breads and other baked goods to brown. This reaction contributes to crust formation and flavor.

makeup and panning   See *shaping*.

marshmallow   A light, foamy confection made from an Italian meringue that is stabilized with gelatin.

marzipan   A thick, sticky, dough-like paste made from ground almonds and sugar and used to coat cakes or in confections.

mascarpone cheese   Mascarpone cheese is known as the Italian cream cheese. It is rich like butter but it has a creamy consistency much like cream cheese.

meniscus   The level of a free-flowing liquid in a measuring container that marks the amount of the liquid the container is holding.

meringue   When egg whites are beaten with sugar to form an egg foam. The ratio of sugar to egg whites depends on what type of meringue is desired. Meringues are used to leaven cakes and soufflés or as a topping on pies or Baked Alaska. They may be baked until crisp and used as a base for cakes and tortes.

metal cake ring   A strip of stainless steel shaped into a circle much like a cake pan with no top or bottom. Metal cake rings are available in various diameters and heights and are used to mold layers of cake with fillings that need to firm up and set before being able to stand on their own. Cake rings can also be used as cake pans by placing them on aluminum foil on a sheet pan and filling them with batter. Also referred to as a *torte ring*.

microplane zester   A long, narrow, rectangular strip of metal, similar to a grater, with raised, sharp cuts, sometimes attached to a handle, used for grating hard cheeses, chocolate, and citrus peels. It is so named after the tool used by carpenters. Also known as a *rasp*.

mise en place (meez ahn plahs)   A French term that means "getting everything ready and in its place" to help a chef get organized by putting all ingredients, tools, and equipment together to get ready to bake.

mixing   The act of combining ingredients.

modeling chocolate   Melted chocolate mixed with corn syrup and kneaded together until a dough-like consistency forms. After several hours, it can be rolled out, cut, and shaped into decorations to top cakes and other pastries.

modified straight dough method   A variation of the straight dough method for mixing yeast breads whereby the fat and the sugar are combined before being added to the other ingredients to ensure their even distribution.

molded cookie   A category of cookie preparation in which a stiff dough is rolled into small balls and baked or flattened with the bottom of a glass, or criss-crossed with a fork and then baked.

mouthfeel   A term used to describe how a food tastes and feels in the mouth.

muffin method   A mixing method in which to prepare quick breads and muffins. The method consists of combining wet ingredients in a bowl before mixing them into a bowl of dry ingredients.

muffin tin   A muffin tin consists of round metal impressions in which muffin batter can be baked to form small cakes or muffins. Muffin tins come in professional sizes (holding 2 to 4 dozen muffins in standard, full, or half sheet pan sizes), miniature (12 muffins), standard (6 muffins), and jumbo
(6 muffins) sizes. Also known as a muffin pan.

natural starter   A mixture of flour, water, and natural or wild yeast that is allowed to ferment. See *sourdough culture*.

neutralization reaction   A chemical reaction between equal amounts of an acid and a base that results in the formation of a salt and water. This results in a neutral pH of 7.0.

nibs   The kernel of the cocoa bean used in the preparation of chocolate.

nutmeg   A spice that originates as a hard seed from the tropical evergreen nutmeg tree. The hard outer coating of the nutmeg seed is covered with a red lace-like material, which when ground, becomes another spice called *mace*. Nutmeg has a strong, sweet aroma used in various sweet and savory dishes.

offset spatula   (1) A spatula with a wide metal blade having a slight bend in the blade just before the handle that is used to remove cookies and small pastries from a sheet pan. (2) A long, round-tipped knife with a slight bend in the blade just before the handle that is used to frost cakes, cookies, or spread fillings. Also referred to as a cake spatula.

old dough   See *pâte fermentée.*

one-stage method   The simplest method of mixing cakes in which all ingredients are added in one bowl, usually dry ingredients first, followed by the gradual addition of liquid ingredients.

osmosis   The tendency of water to go from a higher concentration of water, through a semi-permeable membrane, to a lower concentration of water in an attempt to balance the concentrations and form an equilibrium.

osmotolerant instant active dry yeast   A yeast used in rich sweet doughs. These doughs typically contain greater amounts of sugar, fats, and eggs. Osmotolerant yeast can be rehydrated without large amounts of water present in the dough and without being damaged because of its tolerance of osmotic changes within a dough. See *yeast.*

oven spring   The rapid rising of a yeast dough in the oven as the trapped gases within the dough expand.

overrun   The increase in volume caused by the incorporation of air during the freezing process of ice cream and other churn-frozen desserts.

paddle   A mixing tool used on an electric mixer. Shaped like a boat's oar with open spaces in between, the paddle is used primarily for creaming and blending.

palette knife   A tool similar to an offset spatula but without the bend near the handle. It is used to frost cakes and cookies, and to spread fillings.

parchment paper   Specially treated paper that is used to line cake and sheet pans to prevent foods from sticking. Parchment, also known as baking paper, comes in rectangular sheets $16^{3}\!/_{48}$ by $24^{3}\!/_{48}$ inches (41.5 by 62 cm), which fit perfectly into a full sheet pan. Parchment paper will not burn in a hot oven and can be used to make parchment cones for piping chocolate and thin icings.

passion fruit   A small, round, aromatic tropical fruit with a purple, wrinkled, hard, inedible skin. It has golden flesh with small edible seeds that taste both sweet and tart. The juice can be purchased without seeds in the frozen state and can be used to prepare fillings, mousses, and sauces.

pasteurization   The procedure in which a food substance is heated to a specific temperature for a specific amount of time in order to kill dangerous microorganisms that can cause foodborne illness.

pastry bag   A cone-shaped hollow bag made of various materials and of various sizes with a narrow opening at one end and a large opening at the other end. Frostings, cookie batters, doughs, and chocolate are squeezed through the narrow end of the bag, which is fitted with a pastry tip, to form decorative shapes and designs.

pastry blender   A tool consisting of five to six bent metal wires attached to a handle that is used to cut fat into flour for making pies and quick breads such as biscuits, shortcakes, and scones.

pastry brush   A brush resembling a paintbrush that comes in various sizes and used to lightly cover foods with glazes, butter, water, or egg washes. Also used to brush excess flour off doughs.

pastry cream   A stirred custard consisting of eggs, milk, sugar, and a starch such as flour or cornstarch that is used as a filling for cream pies, fruit tarts, cakes, and cream puffs.

pastry flour   A flour made from soft wheats containing more starch than protein. It is used for more tender pastries and quick breads.

pastry tip   A hollow metal cone shape with varying cuts at the smaller end that is fitted onto a pastry bag such that when frostings or batters are piped out, various designs and shapes are formed. Used to decorate or fill cakes, cookies, and various pastries with frostings, whipped cream, mousses, and pastry cream.

pasteurized fresh whole milk   Whole cow's milk is fortified with vitamin D and nothing artificial. It contains approximately 3.5 percent milkfat and has been heat treated (pasteurized) to 161°F (72°C) for a minimum of 15 seconds to kill bacteria that can cause foodborne illness. Whole milk can be dehydrated and dried to a powder.

pâte à choux (paht uh SHOO)   French for "cabbage paste." A steam-leavened specialty dough used to prepare cream puffs and éclairs that when shaped into small rounds resembles small cabbages. Also known as *éclair paste* or *choux paste.*

pâte brisée (paht bree-ZAY)   French for "broken pastry." Refers to a rich, flaky pastry dough containing flour, fat, and ice water.

pâte fermentée   French for "fermented dough." It is a piece of dough from the previous day's batch of bread dough that is incorporated into the next day's bread dough. Also referred to as *old dough.*

pâte sablée (paht SUB-lay)   French for "sandy dough." The richest and the most tender of the three types of pastry dough. It contains flour, fat, eggs, and more sugar than the other two types of pastry dough (i.e., *pâte brisée* and *pâte sucrée*).

pâte sucrée (paht soo-CRAY)   French for "sugar dough." Contains flour, butter, sugar, and egg yolks. Used as a rich, sweet, pastry dough for fruit tarts, pies, and other pastries.

patent flour   During the milling process, the finest particles of flour taken from the inner part of the endosperm.

pathogen   Microorganisms such as bacteria that cannot be seen by the human eye that can cause foodborne illness.

peanut butter   A ground paste made from peanuts, oil, and salt. By law, in order to be called "peanut butter," the paste must contain at least 90 percent peanuts. Peanut butter is commonly used in many desserts such as confections, cookies, cakes, frostings, and frozen desserts.

pectin   A type of thickener made from the natural sugars within the cell walls of plants, particularly unripened fruits. Pectin is used to thicken jam, jellies, preserves, and fruit glazes.

peel   See *baker's peel.*

persimmon   Small, glossy-skinned fruit resembling a tomato that varies from yellow to red in skin color. It has orange-red flesh with a jelly-like consistency. The two varieties found in the United States are hachiya and fuyu. Persimmons taste sweetest when fully ripened.

piped cookie   A category of cookie preparation in which a dough is pushed through a pastry bag fitted with a plain or decorative pastry tip into various shapes onto a sheet pan and baked. Also called a *pressed cookie.*

pirouette   A type of wafer cookie named after a ballet move. The cookie is rolled up tightly into a cylindrical tube while still warm and pliable.

plasticity   The ability of a fat to hold its solid shape at room temperature while still having the ability to be molded; refers to a fat that has a plastic consistency. An example of a fat with a high degree of plasticity at room temperature is solid vegetable shortening.

pomegranate   A medium, round fruit with a bright pink to red skin encasing an inedible yellowish flesh that is packed with edible seeds that hold a sweet-tart liquid covered with a membrane. Pomegranates are used as garnishes on fruit salads and tarts, and the juice can be purchased separately to prepare various fillings, sauces, and frozen desserts.

poolish   A French word for a thin sponge starter typically prepared with equal parts of flour and water by weight.

popovers   A steam-leavened quick bread made in deep muffin pans forming a puffy brown exterior with a hollow, eggy center.

popover pan   A baking pan similar to a muffin pan but with deeper, narrower impressions with which to bake popovers (puffy, eggy muffin-shaped puffs). The impressions are spaced farther apart to accommodate the rising of the popover.

porous   When a membrane or surface allows air or moisture through it. An eggshell is porous, allowing odors in and moisture to evaporate out.

poured fondant   A cooked sugar syrup that is allowed to crystallize enough to form a sugar paste. It is then melted down and used to glaze baked goods such as napoleons, petit fours, cakes, cookies, and other small pastries.

praline   A caramelized sugar syrup that is poured over nuts (usually almonds or hazelnuts) and allowed to harden. It is then chopped or ground and used to flavor cakes, icings, and candies or to coat cakes and other pastries.

preferment   Means "to ferment before." A mixture of flour, water, and yeast that is allowed to ferment before the actual dough is made. It is then added to other ingredients to form a dough. Preferments provide leavening and flavor to yeast breads.

pressed cookie   See *piped cookie.*

prickly pear   A barrel-shaped fruit from a species of cactus with sharp, prickly thorns. The flesh is a deep purple color. It has small black seeds, similar to a watermelon. Prickly pears have a spongy texture and a mildly sweet flavor. Also known as cactus pears.

profiterole   A small cream puff made from pâte à choux dough.

proof box   A room or cabinet-like box in which humidity and temperature can be controlled to ferment and proof yeast doughs.

proofing   (1) A stage in yeast dough production in which the dough is shaped into rolls, braids, or loaves and allowed to ferment one last time before being baked. (2) A procedure to determine if yeast is alive. The yeast is dissolved in warm water with or without a small amount of sugar. If the mixture becomes foamy after 5 to 10 minutes the yeast is alive and can be used to leaven yeast dough.

protease   An enzyme occurring naturally in dairy products and certain fresh fruits that breaks down proteins. Protease can have a negative effect on gluten formation in yeast doughs and can prevent gelatin from gelling.

protein   Chains or strands of amino acids chemically linked together.

puff pastry   A laminated dough made by enclosing fat into a non-yeasted dough that bakes into a light, flaky pastry with multiple layers leavened by steam.

punching   Also referred to as *degassing*. A stage in yeast dough production in which, after the dough has fermented, the edges of the dough are pulled over into the center to release carbon dioxide and redistribute the yeast.

purifying   Part of the flour milling process whereby air currents are used to blow away any remaining bran pieces on the endosperm after breaking.

quick breads   Refers to a category of breads, scones, biscuits, muffins, and popovers that are quick to make and use chemical leaveners instead of yeast.

quince   A pear-shaped fruit with yellow skin that is always served cooked. It tastes similar to a tart apple or pear.

ramekin   A small baking dish usually made of ceramic or heat-resistant glass used to bake individual soufflés, custards, and cakes. Available in various sizes.

reducing   Part of the milling process whereby endosperm are ground into flour.

refrigerator cookie   A category of cookie preparation in which a stiff dough is shaped into logs, wrapped well, and refrigerated or frozen until firm. The logs are then cut into slices and baked.

resting   A stage in yeast dough production in which, after the dough is scaled and rounded, it is covered and allowed to relax for a short period of time. This allows gluten to relax before shaping. See *benching*.

retarding   A slowing down of the fermentation process by placing dough in a special temperature-controlled box called a retarder or in the refrigerator for several hours.

rhubarb   A long, purplish-pink plant with celery-like stalks used in fillings, cakes, dessert sauces, and pies. The leaves are poisonous and should be carefully trimmed. Rhubarb is usually cooked with sugar to decrease its tartness.

rich dough   A yeast dough that contains greater amounts of fat and sugar. It may also include eggs. The crust of breads made with rich doughs tends to be softer than those made from lean doughs. Some rich yeast breads include brioche, coffee cakes, and cinnamon rolls.

ricotta cheese   Cheese that is made from the reheated liquid whey that is left over after whole cow's milk cheese is made. Curds are formed and then drained. (The Italian variety is made from sheep's milk.) Ricotta is used for such Italian pastries as cheesecake and cannoli.

rolled cookie   A category of cookie preparation whereby a dough is refrigerated until firm and then rolled thin and cut into shapes before baking.

rolled fondant   A cooked sugar paste that is cooled, beaten, and kneaded like a dough. It is rolled out and used to cover cakes and other pastries.

rolled-in doughs   See *laminated doughs.*

rolled-out frostings   Dough-like frostings made up of various ingredients such as fondant, modeling chocolate, or marzipan. Rolled-out frostings can be molded and shaped into decorations or rolled thin to cover cakes and other pastries.

rolling cutter   A tool that resembles a row of small pizza cutters that are joined together. It can expand and contract like an accordion to make different size cuts. It can be used to cut croissant dough, cakes, and cookies.

rounding   A stage in yeast dough production where pieces of yeast dough are rounded into smooth balls, after scaling, forming a smooth, elastic skin of gluten around the outside. Rounding makes the final shaping of the dough easier.

royal icing   A fluffy, uncooked, decorative icing that consists of confectioners' sugar, an acid such as lemon juice or cream of tartar, and egg whites. Resembling a meringue, royal icing dries very hard and is used to create piped decorations or flowers on cakes, cookies, and other small pastries.

rubber spatula   A tool consisting of a soft, rubber, scoop-like spoon attached to a wooden or plastic handle that is used to gently blend together or fold ingredients. Also used to scrape down ingredients from the sides of a mixing bowl and to remove batters and doughs from a spoon.

sabayon   A rich, foamy French custard sauce consisting of egg yolks whisked with sugar or corn syrup over a hot water bath until thickened and a pale yellow. Sometimes white wine is added. The Italian version is known as *zabaglione.*

saffron   A spice from the dried yellow-orange inner threads (the stigma) of the purple crocus plant. It gives foods a beautiful color and it has a pungent aroma. One of the most expensive spices in the world, saffron is handpicked and takes more than 75,000 flowers to produce 1 pound (455 g) of saffron. Besides being used in Middle Eastern and Spanish cuisine, saffron is used to color and scent certain yeast breads and rolls.

Salmonella   A bacteria associated with eggs and poultry that can cause food-borne illness.

saturated fat   Fats directly derived from animals (with the exception of tropical plant oils and cocoa butter) and having a chemical structure wherein there are as many hydrogen atoms as possible bonded to the carbon atoms, and all of the bonds are single. Saturated fats are generally solid at room temperature.

scalding   To heat a liquid to just below the boiling point.

scaling   (1) The act of weighing ingredients. (2) A stage in yeast dough production when the dough is divided and weighed into portions after fermentation and punching.

scoring   See *slashing.*

seizing   When melted chocolate thickens to a dried out, clumpy mass after being exposed to a small quantity of water or moisture.

self-rising flour   All-purpose flour that contains baking powder and salt.

separated egg-foam cake   A type of egg-foam cake in which the eggs are separated and the yolks are beaten with a portion of the sugar until thick and the beaten whites, along with sifted dry ingredients, are folded in alternately.

shaping   A stage in yeast dough production in which scaled pieces of dough are formed into the desired shapes that will be placed into the oven. Also called *makeup and panning*.

sheet cookie   The category of cookie preparation in which a batter is spread into a sheet pan with sides and baked. The baked sheet is then cut into squares or other shapes.

sheet pans   Sheet pans are rectangular metal baking pans. They come in two sizes: full and half. The full sheet pan measures 18 by 26 inches (45 by 65 cm) and has sides that are 1 inch (2.5 cm) high. The half sheet pan measures 13 by 18 inches (32.5 by 45 cm). They are so named because two half sheet pans put together would equal one full sheet pan. Full sheet pans are used mostly in commercial kitchens because they are too large to fit into a standard sized noncommercial oven. Most half sheet pans fit into noncommercial ovens.

sieve   A small tool used to separate finer particles from coarser ones. It consists of a handle attached to a metal bowl with screen-like openings. The sieve is used to sift out lumps from dry ingredients like flour or confectioners' sugar, to strain lumps from desserts like pastry cream or custard sauces, or to separate out solids from liquid ingredients. Also known as a *strainer*.

sifter   A metal cup with a screened bottom and rotating metal wires inside to help separate out lumps in dry ingredients like flour, confectioners' sugar, and cocoa powder. Used for dry ingredients only, the sifter resembles a sieve.

sifting   (1) Part of the flour milling process whereby particles of flour are sorted by size. (2) The act of putting dry ingredients through a sifter.

silicone baking mats   Reusable, flexible, plastic rectangular sheets coated with silicone and able to withstand extreme temperatures, both cold and hot. Resembling rubber placemats, they are placed into full sheet or half sheet pans instead of parchment paper with no need to grease or flour them. Wiped clean with a damp sponge or cloth, silicone baking mats are reusable.

simple icing   An easy icing consisting of confectioners' sugar and water, cream, milk, citrus juice, or corn syrup added with flavorings to make a thin, pourable icing that can be drizzled over coffee cakes, scones, cookies, or sweet yeast breads.

simple syrup   A sugar syrup consisting of equal parts of water and granulated sugar by weight that is brought to a boil until the sugar is dissolved. Simple syrups can be flavored and brushed onto cake layers to keep them moist, used in a base for frozen desserts, or brushed onto the tops of pastries as a glaze.

single-acting baking powder   A chemical leavener that requires moisture to produce carbon dioxide gas to leaven baked goods. It is rarely used because all leavening power would be gone as soon as the ingredients were moistened.

skim milk   Whole cow's milk that has had most of the milkfat removed so that it contains up to 0.5 percent milkfat. Also referred to as *fat-free* or *nonfat milk*.

slashing   Shallow cuts made with a razor on the surface of an unbaked loaf of yeast bread to allow the bread to expand during baking. The cuts may also be decorative. Also referred to as *scoring*.

smoked gouda   A mild cheese made from cow's milk originating from Holland that has been exposed to smoke, giving it a brownish color and smoky flavor.

soft peaks   The stage of beating a meringue until the beater or whip is held up and the meringue curls over on itself.

soft wheat   Wheat that is grown in milder climates and contains less protein and a higher starch content.

solute   The substance that is dissolved in a solution.

solution   A *solute* and *solvent* that are evenly distributed when mixed together.

solvent   The liquid that a substance is placed in to help it to dissolve.

soufflé   A light, airy French dish consisting of a base for structure and added flavorings, to which beaten egg whites are folded in to provide leavening. Soufflés can be baked or frozen in small ceramic dishes called *ramekins* or soufflé cups.

sourdough culture   A live, bubbly mixture used to leaven bread doughs that uses flour, water, and wild or natural yeast and bacteria that produce flavorful by-products through the process of fermentation. Breads baked with sourdough cultures have a tangy, acidic flavor. See *natural starter* and *starter.*

sourdoughs   A type of preferment that uses a natural or wild yeast starter. Sourdough starters are used to leaven bread doughs. Unused starter can be replenished with flour and water and maintained for long periods of time.

sour cream   Cream that has been soured or fermented with harmless bacteria to give it a thicker consistency and a tangy flavor. It may contain other ingredients such as gelatin or enzymes to help it thicken. It contains between 18 and 20 percent milkfat.

spices   Seasonings derived from aromatic dried plants containing essential oils that impart deep flavor to foods. The various parts of a plant that can be used for spices include the bark, flower buds, berries, seeds, and roots. They are available whole or ground. The quality of a spice depends on the method of harvesting, processing, and climatic conditions.

spiking   When a small amount of commercial yeast is added to a natural starter to create more leavening power.

sponge   A type of preferment consisting of a mixture of flour, water, and yeast that is allowed to ferment for 30 minutes to several hours before other ingredients are added to make a yeast dough. Sponges add flavor to yeast doughs and give the leavening process a head start.

sponge method   A method of yeast dough preparation in which the dough is prepared in two stages. A portion of the flour, water, and yeast are mixed together and allowed to ferment before other ingredients are added to make a dough. The sponge or "pre-dough" is referred to as a preferment and imparts great flavor and leavening power to yeast breads.

springform pan   A springform pan is a round pan of variable size with a removable bottom. It resembles a cake pan but has higher sides for baking cheesecakes and other cakes and desserts. The sides of the pan can be detached from the bottom for ease of removal.

star fruit   A fruit native to Asia and resembling a five-pointed star when cut crosswise. The taste ranges from tart to sweet, similar to a mild Granny Smith apple. Also known as *carambola*.

starch retrogradation   The process of staling whereby chemical changes occur in the molecular structure of a baked good causing the starches to bond more closely over time, forming a drier texture. Also known as chemical staling.

starches   Long chains of sugars chemically bonded to one another in the form of semicrystalline shapes. Referred to as starch granules, the inside structure consists of amylose and amylopectin. Used to thicken desserts such as pastry creams and pie fillings.

starter   A mixture of flour, water, and yeast (either commercial or wild) that is allowed to ferment before a portion of it is added to other ingredients to make a yeast dough. The remaining starter is saved and fed additional flour and water to maintain it over a period of time until it is used again. Starters help leaven yeast doughs and impart complex flavors to yeast breads.

stiff peaks   The stage of beating a meringue until the beater or whip is held up and the meringue stands straight up in a vertical peak.

still-frozen desserts   Frozen desserts that lay still in the freezer with no agitation or churning as they are being frozen. Still-frozen desserts include mousses, semifreddos, bombes, and parfaits.

stippling   See *docking.*

stirred custard   A mixture of eggs and milk or cream stirred on top of the stove with or without the addition of a starch and cooked until thickened.

straight dough   A yeast dough in which most of the ingredients are mixed together in one bowl.

straight dough method   The simplest mixing method for yeast breads in which all the ingredients are added in one bowl and mixed.

straight flour   The particles of flour from the entire endosperm.

straight spatula   A long, rounded metal knife used to spread frostings and fillings. Also known as a *palette knife.*

streams   Particles of milled flour that are sorted by size in order to be classified.

sucrose   The chemical name for table or granulated sugar; derived from the sugar cane or sugar beet plant. Sucrose is composed of two simple sugars—glucose and fructose—bound as one molecule.

sugar bloom   A recrystallization of sugar that forms on the surface of chocolate that has been exposed to moisture giving it a whitish coating.

sugar syrup   When one or more sugars are dissolved in water. Sugar syrups cooked to specific temperature are the foundation of many desserts including caramel, Italian meringues, marshmallows, fudge, and pulled sugar decorations.

surface tension   The natural tendency for two immiscible (not mixable) liquids to separate.

suspension   Tiny liquid drops or solid particles floating freely in another liquid.

sweetened condensed milk   Whole milk that has 60 percent of the water evaporated from it and with extra sugar added, leaving a sweet, thickened liquid that is sold in cans. Sweetened condensed milk is used in desserts such as confections, cheesecakes, and custards.

Swiss meringue   A type of meringue in which egg whites are warmed with sugar over simmering water before being beaten.

temperature danger zone   The temperature range between 41° and 135°F (5° and 57°C) in which bacteria grows very quickly in foods. This temperature range may vary from state to state.

tempering   (1) The act of bringing eggs up to the proper temperature by slowly whisking in a hot liquid (e.g., hot milk) to prevent curdling. (2) The stabilization of fat crystals within chocolate through a heating and cooling process.

thickener   A food that helps ingredients to become less fluid and more dense.

three-fold or **letterfold turn**   When a laminated dough is folded into thirds like a letter to incorporate the fat and to produce multiple layers of fat and dough.

torte ring   See *metal cake ring*.

trans-fats   Liquid fats such as oils that are partially hydrogenated to a solid or partially solid state at room temperature. The chemical structure of the fat is such that the hydrogen atoms sit diagonally across the double bond. Trans-fats are associated with an increased risk of heart disease and certain cancers.

truffles   A rich chocolate confection prepared from a mixture of heavy cream and chocolate (ganache) flavored with extracts, alcohol, fruits, nuts, or purées. The mixture is chilled and rolled into balls where they can be served as is or dipped into melted chocolate.

tube pan   A deep, round cake pan with a hollow tube in its center used to bake angel food, sponge, and pound cakes. Some tube pans come with a false bottom or have metal tabs that stick out from the top edge of the pan to allow cakes to balance upside down while cooling.

tuile (tweel)   French for "roof tile." A type of crisp, thin *wafer cookie* that can be molded into various shapes while warm before cooling and hardening.

tunneling   The result of too much gluten produced by overmixing that results in large holes or cavities "tunneling" through the inside of a quick bread.

turn   The sequence of rolling out and folding a laminated dough to create multiple layers.

turntable   A rotating, elevated plate used to facilitate frosting a cake.

two-stage method   A cake mixing method in which the liquids are added in two stages. High-fat cakes that contain more sugar by weight than flour (high-ratio cakes) are mixed using this method.

ultrapasteurization   When milk or cream is held at an even higher temperature (280° to 300°F; 138° to 150°C) than regular pasteurization requires. The product has a longer shelf life but the cream will not attain the same volume when whipped. Also known as ultrahigh-temperature pasteurization (UHT).

unsaturated fat   A type of fat derived from plants that tends to be liquid at room temperature (with the exception of partially hydrogenated vegetable shortenings). In the chemical structure, fewer than the maximum number of hydrogen atoms are bonded to each carbon atom, creating some double bonds between carbon atoms.

unsweetened chocolate   Chocolate liquor that has been cooled and molded into bricks or disks. Also referred to as *baking chocolate* or bitter chocolate.

wafer cookie   A category of cookie preparation in which a batter is spread onto a sheet pan or over a stencil before being baked. Wafer cookies are soft when hot and harden as they cool. See *tuile,* a type of wafer cookie.

wash   A liquid used to brush over food such as a yeast dough before baking to add color or shine, or to help toppings adhere.

wet method   A method of preparing caramelized sugar by heating sugar and water and boiling it to the caramel stage.

wheat berry   The whole wheat kernel before processing or milling consisting of the *hull* or *bran, endosperm,* and *germ.*

wheat kernel   See *wheat berry.*

whip   A tool used on an electric mixer that is shaped like a wire whisk and used to beat air into eggs or cream.

whisk   A tool having thin, curved metal loops attached to a main handle. Whisks are used to combine dry or wet ingredients and to whip air into ingredients such as egg whites or heavy cream.

white chocolate   Consists of cocoa butter, milk solids, sugar, and flavorings, but because it contains no chocolate liquor, it is not really chocolate. There are two qualities of white chocolate: those that contain cocoa butter and are of a higher quality and those that contain little or no cocoa butter.

whole egg-foam cake   A type of egg-foam cake in which whole eggs are beaten with sugar and then sifted dry ingredients are folded in.

whole milk   Milk from a cow containing approximately 3.5 percent milkfat.

whole wheat flour   Flour that is milled from all three components of the *wheat kernel*: the *bran* or *hull*; the *germ*; and the *endosperm.*

yeast   A one-celled microscopic living fungus that undergoes fermentation and that is used as a leavening agent for yeast breads and in the production of cheese, beer, and wine. Available as *fresh* or *compressed*, active dry, *instant active dry*, and *osmotolerant instant active dry*.

yogurt   Milk that is heated with special bacteria until it ferments and becomes thick and tangy. The fat content varies depending on how much butterfat the milk contains. Yogurt can be purchased as plain and unsweetened, which is best used for baking, or sweetened and flavored, which can be eaten on its own. Mildly flavored yogurts (e.g., vanilla and lemon) can be easily substituted for sour cream in baked goods.

zabaglione (zah-bahl-YOH-nay)   The Italian version of a sabayon, a sweet custard sauce that uses sweet Marsala wine. See *sabayon*.

zester   A tool consisting of five, small, angled holes with sharp edges attached to a handle used to remove long, narrow strips from the outermost peel or rind of citrus fruits.